A Shock to the System

Restructuring America's Electricity Industry

A Shock to the System

Restructuring America's Electricity Industry

by
Timothy J. Brennan
Karen L. Palmer
Raymond J. Kopp
Alan J. Krupnick
Vito Stagliano
and
Dallas Burtraw

Resources for the Future
New York

Published 1996 by Resources for the Future
2 Park Square, Milton Park, Abingdon, Oxon, OX14 4RN
711 Third Avenue, New York, NY 10017

Library of Congress Cataloging-in-Publication Data

A shock to the system : restructuring America's electricity industry / by Timothy
    J. Brennan ... [et al.].
        p.   cm.
    Includes index.
    ISBN 0–915707–80–2 (pbk. : alk. paper)
    1. Electric utilities—Government policy—United states. 2. Energy pol-
icy—United states. I. Brennan, Timothy J.
HD9685.U5S45 1996
333.79′32′0973–dc20   96–20204
                                                                    CIP

This book was edited by Betsy Kulamer and designed by Diane Kelly, Kelly Design.
Photos on pages 8 and 70 are printed with permission. Copyright 1996 PhotoDisk, Inc.

RESOURCES FOR THE FUTURE (RFF) is an independent nonprofit organization engaged in research and public education on natural resources and environmental issues. Its mission is to create and disseminate knowledge that helps people make better decisions about the conservation and use of their natural resources and the environment. RFF takes responsibility for the selection of subjects for study and for the appointment of fellows, as well as for their freedom of inquiry. RFF neither lobbies nor takes positions on current policy issues.

Because the work of RFF focuses on how people make use of scarce resources, its primary research discipline is economics. Supplementary research disciplines include ecology, engineering, operations research, and geography, as well as many other social sciences. Staff members pursue a wide variety of interests, including the environmental effects of transportation, environmental protection and economic development, Superfund, forest economics, recycling, environmental equity, the costs and benefits of pollution control, energy, law and economics, and quantitative risk assessment.

Acting on the conviction that good research and policy analysis must be put into service to be truly useful, RFF communicates its findings to government and industry officials, public interest advocacy groups, nonprofit organizations, academic researchers, and the press. It produces a range of publications and sponsors conferences, seminars, workshops, and briefings. Staff members write articles for journals, magazines, and newspapers, provide expert testimony, and serve on public and private advisory committees. The views they express are in all cases their own and do not represent positions held by RFF, its officers, or trustees.

Established in 1952, RFF derives its operating budget in approximately equal amounts from three sources: investment income from a reserve fund; government grants; and contributions from corporations, foundations, and individuals. (Corporate support cannot be earmarked for specific research projects.) Some 45 percent of RFF's total funding is unrestricted and provides crucial support for its foundational research and outreach and educational operations. RFF is a publicly funded organization under Section 501(c)(3) of the Internal Revenue Code, and all contributions to its work are tax deductible.

Contents

List of Figures and Boxes

ix

Foreword

Since the oil crises of the 1970s and the deregulation of natural gas in the 1980s, energy markets around the world have been generally quiet—with one major exception. That major exception, of course, is the electricity industry. The rapid and very striking changes that the industry currently is undergoing are the subject of this book.

In a nutshell, the electricity industry is facing the same forces of competition that transformed the transportation, financial, and telephone industries in the past two decades. Except for very recent experiments in a few states, industrial, commercial, and residential consumers of electricity have had little choice but to accept the power provided by their local regulated electric utility company at prices established by their state public utility commission. Not that long ago, those local utilities were unable to take advantage of competition among electricity generators to get power at the lowest possible prices. Technological change, as well as recent legislative and regulatory initiatives at both the federal and state levels, has made competition in power generation possible. The day may not be far off when even residential electricity consumers—people like you and me—will be able to shop for bargain prices and service options.

But getting from where we once were to where we will end up is by no means straightforward. Important decisions must be made at many steps along the road to a much more competitive market for electricity. And these decisions will determine in part the prices that eventually will be paid, the quality of service that all customers will enjoy, the environmental consequences of electricity production and use, and the profits to be earned by investors in the companies that provide electricity or that use large amounts of it. In fact, given the large and growing importance of electricity in the U.S. economy, it is no exaggeration to say that the way we manage the transition to a more competitive electricity industry will help determine the competitiveness of U.S. firms in international markets. Clearly, the stakes are large.

In this book, Tim Brennan, Karen Palmer, and the other contributors to *A Shock to the System* present no brief for a particular set of answers to the important questions facing the country. Rather, they describe how the electricity industry got to where it finds itself in mid-1996, what major choices face the many parties (including regulators and elected and appointed officials at all levels of government, as well as those in the business community and public interest groups) that will play a role in answering the leading questions, and what are—in the authors' views—the pros and cons associated with each of the leading options. *A Shock to the System*, then, is ideally suited to those searching for an independent and unbiased introduction to a truly important public policy issue.

This book is more than the culmination of much hard work on the part of the authors. It also marks, we hope, the initiation of an extended program of research at Resources for the Future on issues related to the restructuring of the electric utility industry in the United States and around the world. Having invested in what economists call the "human capital" (or "smarts" to the noneconomist) necessary to produce this book, we intend to use what we have learned to further enlighten the debate. It is our hope and expectation that you will benefit from this book. If so, please be on the lookout for the additional research, policy analysis, and outreach from us.

Paul R. Portney
President
Resources for the Future

Preface

Blaise Pascal once apologized to a correspondent for writing a long letter, claiming that he "lacked the time to make it short." Having spent the better part of the last eighteen months attempting to boil down the overwhelming cauldron of the electricity competition policy debate to a digestible stew, we have some idea what Mr. Pascal meant.

In early 1995, we at Resources for the Future (RFF), along with many others, saw that the electricity industry was beginning to move away from a traditionally regulated and closed system and toward a much less regulated and more competitive market. Many policymakers and the public were caught by surprise when similar changes took place in the airline, banking, and telecommunications industries during the 1980s. To keep that from happening again, we prepared a series of papers on the history of regulation in electricity, proposals for implementing competition, recovery of so-called "stranded costs," and the projected environmental consequences of restructuring.

In April 1995, we presented these papers to a gathering of experts from industry, public interest groups, academia, and the policy community, held in conjunction with the annual RFF Council meeting. We took advantage of the meeting's location in California to exploit the pioneering efforts in that state to expand electricity competition. Formal comments were prepared by Ron Russell, then-commissioner on the Michigan Public Service Commission; Roger Naill, vice president of Applied Energy Services, Inc.; Kathy Treleven, senior strategic planner at Pacific Gas and Electric Company; and Ralph Cavanagh, energy program director for utility works of the Natural Resources Defense Council. Armed with the comments from these discussants and the other participants, we completed the initial draft of this book.

To distill the vast and complex historical, economic, regulatory, technical, and environmental aspects of the electricity industry, we assembled a diverse team with a wide range of experience. While we all formally and informally contributed to the book as a whole, usually some of us had primary responsibility for specific chapters. Karen Palmer and Ray Kopp, respectively fellow in and director of the Quality of the Environment Division at RFF, prepared the introduction and summary in Chapter 1. The historical and institutional perspectives in Chapter 2 were contributed by Vito Stagliano, a visiting fellow at RFF and formerly assistant secretary for policy at the U.S. Department of Energy.

The task in Chapter 3 of interpreting and summarizing the multifaceted proposals for implementing wholesale and retail competition was taken on by Karen Palmer, drawing on her experience in state regulatory policy and the economics of electricity markets. Working with Ray and Karen, Tim Brennan made use of his background in regulatory economics and industrial organization to draft Chapters 4 through 6 on transmission regulation, industry restructuring, and stranded-cost recovery. Tim was a Gilbert White Fellow at RFF during 1995 and is now a senior fellow there, as well as a professor of policy sciences and economics with the University of Maryland. Alan Krupnick and Dallas Burtraw, senior fellow and fellow respectively in RFF's Quality of the Environment Division, brought their expertise in the evaluation of pollution policies to preparing the chapter on environmental consequences.

The contributions RFF made to this project go beyond the listed authors. Our largest debt of gratitude goes to RFF President Paul R. Portney, whose careful and extensive questions and comments were indispensable in helping to make our work clear and useful. His commitment to the project, along with that of Ted Hand, RFF's vice president for finance and administration, was instrumental in completing this project and in making it useful to the energy policy community. Douglas R. Bohi, director of RFF's Energy and Natural Resources Division, lent us his extensive knowledge of the electricity industry and of how federal and state governments are struggling to guide and adapt to increasing competition and industry restructuring. We are also grateful to James Boyd, whose analyses of stranded costs and of the peculiar ways in which the physics of electricity transmission affects the economics of that business did a great deal to help us understand these issues. Last and not least is our appreciation for the expert research assistance we received from Kristi Mahrt, Tim Vandenburg, and Brian Kropp, and the thorough and thoughtful editing by Betsy Kulamer.

Late in the summer of 1995, we circulated a draft of this book to a number of industry experts from across the country to improve further our understanding of the technical, financial, and legal factors affecting the restructuring controversy. We received feedback from Frederick E. John, James R. Hendricks, William

McCormick, Dale Landgren, Susan Chilson, Alyson B. Huey, Jackie Pfannenstiel, Kathy Treleven, Tony Borden, John Rowe, Linda Stuntz, and Everard Munsey. The book reflects suggestions and corrections from all of these sources and is very much the better for them. Our gratitude for the help of the participants at the 1995 RFF Council meeting, our colleagues at RFF, and these reviewers impels us to ensure that readers know that we bear sole responsibility for all errors and omissions.

Many times, presenting policy debates in an unbiased way is a difficult task, as the phrase "bending over backwards to be fair" exemplifies. In this case, however, the uncertainty is so large, the stakes so great, and the feeling on different sides of the issue so intense, that fairness may be among the easiest of our goals to achieve. Readers who wish we took a stronger stand on one side of an issue can rest assured that others believe we should have taken the stronger stand on the other side.

Our belief at the outset of this project, and one that we hold as it comes to completion, is that the technical, economic, and political questions that will determine how electricity restructuring evolves are yet to be answered. Our hope is that this volume will encourage participants in these debates to ask the right questions and demand the right answers. Our objective is to increase the likelihood that policies toward the electricity industry of the future are those supported by the best clear and objective research and analysis, and not merely the outcome of political bargains.

Tim Brennan and Karen Palmer
Resources for the Future

Key to Icons

Throughout this book, icons are used to represent functions or participants in the electricity industry, now and as it may be restructured. The following is a key to the icons and what they represent.

 Generation companies

 Transmission network (grid)

 Independent system operator (ISO)

 Poolco

 Local distribution system

 Line company

 Marketing company

 Final customers: Residential

 Final customers: Commercial

Final customers: Industrial

Chapter 1

Introduction
Why Care about Restructuring the Electricity Industry?

Electricity is perhaps the most common item consumed in the United States and the developed world. In 1994, retail electricity sales in the United States topped $200 billion, more than we spent as a nation on automobiles, telecommunications, or colleges and universities (see figure on page 5). Electricity use pervades all facets of our daily life—at home, at work, and at play. Yet few of us pay much attention to how electricity is produced, who is responsible for its production, how it comes to be delivered where we need it, and how the prices we pay for it are determined. Only when we suffer the occasional power outage do we realize how much we depend on electricity and how much we take for granted the complex system that brings it to our homes, offices, and factories.

1

This book is about the U.S. electricity industry and how it may change over the next decade. Major proposed reform of federal and state regulations, including deregulation for some segments of the industry, is intended to increase competition in the production and sale of electricity. As these proposals would affect the internal operation and external organization of the utilities that have traditionally provided most of our electricity, these reforms are generally referred to as *restructuring* the electricity industry. Pointing to the effects of the deregulation of airlines in the 1970s and of long-distance telephone service in the 1980s, many analysts hope that greater competition in the electricity industry will give rise to lower rates and expanded forms of service.

Greater competition in the electricity industry may lead to lower rates and expanded forms of service, but it also may require that consumers become more sophisticated and that the reliability of the electricity system be ensured.

But alongside these potential benefits, expanding competition in the electricity industry also introduces complications and raises some concerns. For example, under some reform plans, individual households would be able to purchase their electricity from one of several possible suppliers. In this case, as when long-distance telephone service was deregulated, customers used to purchasing a standard service from the only available supplier may need to become more knowledgeable if they are to choose the electricity supplier, the package of services, and the associated prices that best meet their needs. Under a different scenario, participation in competitive electricity markets may be limited to large consumers of electricity (factories and office buildings, say), in which case households may see no benefit from regulatory reform. They may even end up paying *higher* prices for electricity if they come to bear a greater share of the fixed costs of existing electric utilities. The continued reliability of the electricity system also may become a greater concern as the industry responds to diminished government oversight.

The potential benefits of bringing more competition to the electricity industry—lower prices, reduced production costs, more services—will depend on how competition is put into effect. Will it stop at the wholesale level, where utilities and energy marketers are the only buyers? Or, will competition filter down to households and businesses? What segments of the industry will continue to be regulated and what rules will apply? Different answers to these and other questions are found in the many proposals offered by state and federal regulators, existing utilities, customer organizations, and other interested parties. The range of proposals and, indeed, the substance of individual proposals keep changing as the debate unfolds.

Rather than attempt to take a snapshot of this fast-moving train, our purpose in this book is to give readers the background they need to understand and evaluate the ever-changing array of proposals for introducing competition to electricity markets. We introduce readers to the crucial elements, concepts, and industry terminology used in discussions about restructuring in the electricity industry. In the following chapters, we discuss the significant issues in the debate about competition, focusing on the consequences that the major proposals would have on efficiency, market structure, regulation, and the environment. While making policy inevitably takes into account winners and losers, we approach the issues and the proposals from an economic standpoint. This means that we give center stage to considering the impact that the resulting industrial structure would have on the efficiency of the economy as a whole—including the recognition and control of environmental costs.

As we review the central topics in the debate over restructuring the electricity industry, we hope to indicate not just the current state of knowledge but, more importantly, the questions policymakers must answer if the benefits of competition are to be realized with a minimum number of disruptions along the way. At the time of writing, we believe that the debate is still in its early stages, with many crucial issues unsettled. The magnitude of the political and financial stakes inevitably produces strong convictions by various parties about the matters at hand.

The need for more research is widespread and growing. Only time and experience will resolve many of the questions regarding implementation, regulation, the makeup of the utility industry, cost recovery, and environmental effects. We hope that this book fairly and usefully frames the issues and describes the factors that policymakers, industry members, customers, and others concerned with the public interest should consider in predicting how competition in the industry *will* develop and in forming judgments about how that competition *should* develop.

The Coming of Competition to the Electricity Industry

Most observers of the electricity industry segment it by its three functions: power *generation*, long-distance *transmission*, and local *distribution*. (For a more detailed description of these functions, see Chapter 2.) All three of these functions usually have been integrated in a single entity—the traditional electric utility.

In 1907, states began to regulate the rates that utilities could charge their retail customers and, in 1935, the federal government began to regulate wholesale electricity rates. In both cases, regulated rates were set high enough to ensure that utility investors had the opportunity to earn "a fair rate of return" on their investments, but no more than that. Regulation of the electricity industry was justified on the grounds that electricity generation, transmission, and distribution were *natural monopolies*—in other words, the cost of generating, transmitting, or distributing electricity would be lower if only one firm undertook each activity and competition could not be relied upon to protect customers from monopoly pricing. (For more about natural monopolies, see box in Chapter 4 on page 65.)

During the past several years, traditional economic regulation of the electricity industry has come to appear no longer necessary or desirable. As technology in the industry has progressed, electricity production no longer fits the natural monopoly mold. Small, modularized generation systems can be manufactured and shipped to locations where either they could be plugged into the existing transmission system or they could provide electricity to a single consuming unit, such as a large factory. Moreover, these new, smaller units can generate electricity at the same low cost as the very large central power stations that were built only a few decades ago.

During this same period, the perception arose that economic regulation of electric utilities was failing in its mission. By basing prices on costs, the traditional approach to regulation severely limited incentives for utilities to reduce costs, since such efforts would lead to commensurate reductions in revenues and profits. In addition, basing prices on costs produced substantial variation in retail electricity prices across the country, with many high-cost utilities charging prices more than twice those charged by low-cost utilities. Shielded from competition within their franchised service territories, regulated utilities have felt little pressure to adapt or expand their service offerings to meet changing customer needs. Motivated by federal legislation in the late 1970s, utility regulators in several states have been requiring the utilities they regulate to purchase power from third-party generators—known as *nonutility generators*—if these generators agree to sell power at a sufficiently low price and to meet certain other conditions. While this practice has introduced some competition into wholesale electricity markets in several states, it has not always led to lower prices for electricity customers.

To remedy this regulatory failure and to enhance the efficiency of the electricity industry, some observers suggested introducing competition to the generation function and developing new approaches to regulating the remaining natural monopolies in the transmission and distribution functions. Proposals to deregulate electricity generation and expand competitive electric power markets are currently under consideration by state and federal regulators, and these proposals offer a variety of approaches. In some, owners of newly deregulated generating capacity could sell electricity directly to utilities at prices determined solely by market

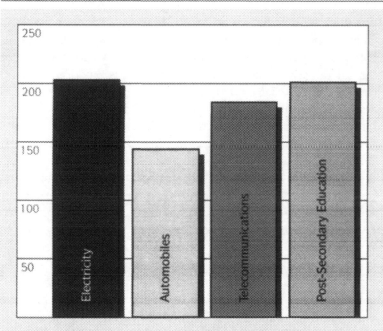

Retail sales of electricity in the United States relative to other nationwide expenditures in 1994 (millions of dollars)

Sources: **Electricity:** U.S. Department of Energy, *Electric Sales and Revenue 1994,* 1994; **Automobiles:** U.S. Department of Commerce, *Survey of Current Business,* 1995; **Telecommunications:** Federal Communications Commission, *Telecommunications Industry Revenue: TRS Fund Worksheet Data,* 1996; **Post-Secondary Education:** U.S. Department of Education, *Digest of Education Statistics 1995,* 1995.

forces; the utilities would then transmit and distribute the electricity to industrial, commercial, and residential customers. In more expansive proposals, large industrial customers for electricity could contract directly with a generator for the provision of power at prices negotiated by the two parties. As these markets develop, eventually residential customers may buy electricity directly from generators, much the same way they do now for long-distance telephone service. Each of these proposals and all their variants depend on the manner in which existing regulations are relaxed or reformed at both the state and federal levels, as well as on the form of the competitive electricity market that is allowed to develop.

Policy Initiatives Shaping Regulatory Reform

As we have suggested already, the range of state and federal initiatives to promote, evaluate, and respond to increasing competition in the electricity industry is complex and constantly changing. Ideas and proposals rise, fall, and rise again. Not only are there disputes about what to do, there is wide disagreement about whether the authority to expand competition and to be responsible for the effects of expanded competition lies with the states or the federal government. Given the continuous flux of most policy proposals in this area, a comprehensive summary of current federal and state initiatives would be quickly out of date. However, a recapitulation of the prominent proposals and of the debates surrounding them can illustrate the range and complexity of the issues involved.

Discussions regarding all the proposals for expanding competition in the electricity industry often refer to the coming restructuring of the industry. Within the context of this debate, *restructuring* refers to changes in the ownership or internal operation of the utility that aim to bring about the separation of the generation, transmission, and distribution functions. While the exact nature of the changes is yet to be defined, policymakers generally agree that some degree of separation of these functions would be a prerequisite for competition. Often, either the term "restructuring" or the phrase "restructuring of the electricity industry" is used to denote the broader concept of movement toward increased competition in the industry.

Federal Initiatives

The chief regulatory initiative by the federal government comes in response to requirements set forth by Congress in the Energy Policy Act of 1992. Since 1977, the primary federal agency responsible for regulating electricity has been the Federal Energy Regulatory Commission (FERC). Recognizing that access to the electricity system is essential if competitive generation markets are to evolve, Congress—via the Energy Policy Act—required that FERC force transmission-owning utilities to deliver power from generators to other utilities and electricity wholesalers at reasonable, nondiscriminatory, cost-based rates. To carry out this mandate, in April 1996 FERC issued Order 888, which sets out procedures for implementing this requirement. FERC also tackled the issue of compensating utilities for the substantial economic losses that are likely to result from transition to a competitive marketplace. We consider this FERC Order in more detail in Chapters 4 and 6.

State Initiatives

Many states also are formulating proposals to expand competition in electricity markets. In 1995, regulators in twenty-nine states and the District of Columbia formally considered the issues of how and whether to expand competition in the electricity markets under their jurisdictions. Most notable among the twenty-nine states was California, where, after more than a year of deliberation, state regulators issued a plan in late December 1995 that proposes to allow all customers, including residential customers, to contract directly with electricity suppliers by January 1, 2003.

Ideas and proposals to restructure the electricity industry rise, fall, and rise again. Not only are there disputes about what to do; analysts disagree about whether states or the federal government have authority to expand competition.

Two of the most controversial aspects of the California plan are, first, that utilities will be allowed to recover all of their "stranded" costs of past investments and of existing contracts made uneconomical by the introduction of competition and, second, that utilities will be required to buy and sell electricity in a government-mandated, real-time electricity market. The California plan also proposes a new, complicated, and untested approach to pricing electricity transmission based on differences in the prices of electricity generated at different locations on the transmission grid. Portions of the California plan are subject to approval by the state legislature and by FERC, so more debate about it is certainly forthcoming.

Jurisdictional Issues

The array of proposals for expanding competition in the electricity industry brings to light unresolved issues about who has authority to do what, especially when it comes to extending competition to the retail level.

Chief among these issues is which regulatory body—if any—has the authority to implement retail competition. While the Energy Policy Act of 1992 mandates access to the transmission grid for deliveries to utilities and wholesalers, it also prohibits FERC from forcing utilities to transmit power for another generator that is selling directly to a retail customer. Whether state regulators have the authority to order transmission access to retail customers has not been determined. Congressional action ultimately may be necessary to implement full retail competition. In addition, some members of Congress have expressed interest in a national approach to restructuring of electricity markets, perhaps similar to recent legislation designed to bring competition to local telecommunications markets. One argument for not proceeding on a state-by-state basis is that the boundaries of regional electricity markets rarely correspond to state borders and, therefore, cooperation among states may be desirable.

Another aspect of this jurisdictional issue involves identifying the dividing line between transmission, which is regulated by FERC, and distribution, which is regulated by the states. The distinction becomes crucial if utilities are going to sell transmission

Local power lines bring electricity to our homes, offices, and factories, yet few of us pay much attention to how it is made, who makes it, who delivers it, or how its price is set.

and distribution services separately to independent, retail electricity providers and customers. While FERC offers some guidelines for distinguishing between the two types of service in its 1996 transmission access rules, this issue is far from resolved.

Policy Initiatives in Other Countries

The movement to bring competition to the electricity industry extends far beyond the United States. Indeed, in many countries (including the United Kingdom, Norway, Argentina, Chile, and New Zealand), the process has advanced much further than it has in the United States. For example, in the early 1990s, the British government simultaneously privatized and disintegrated the publicly owned electricity industry supplying England and Wales into several separate generation companies, several regional distribution companies, and a single transmission company. The transformed British system also includes a government-mandated electricity spot market that coordinates short-term electricity trades, as well as operation of the transmission grid. By 1998, all retail electricity customers will be permitted to pick their own electricity suppliers.

As part of this transformation, the United Kingdom has imposed price-cap regulation on regional electricity distributors. This method of regulation, which caps the rate of growth in the price of electricity according to a prespecified formula, is being considered for adoption in several states in the United States. The successes and failures of the institutions adopted to facilitate competition in generation and to implement regulation of transmission and distribution in the United Kingdom and other countries should provide some lessons for U.S. policymakers.

The Six Major Issues

To help readers grasp the matters that lie at the heart of the competition and restructuring debate, we focus on six issues:

1. the current structure of the electricity industry and the accumulated history of legislation and regulation affecting the industry;
2. the organization of the new competitive markets that will match buyers with sellers of electricity;
3. regulation and pricing of transmission services;
4. the effect on the efficiency of the industry of separating generation from transmission and distribution;
5. the economic and politically sensitive issue of stranded costs, that is, the losses that utilities may bear as a result of the implementation of a competitive market; and
6. the implications of competition and restructuring for environmental quality, renewable energy sources, and conservation.

Each of these issues occupies a chapter of this book, and the following overview briefly summarizes the issues discussed in more detail within the chapters.

Current Structure and History

The current structure and performance of the U.S. electricity industry is the result of nearly ninety years of government regulation and countless pieces of legislation designed to improve the system or meet some political goal. A brief review of this history is helpful in understanding how the past has shaped the current crop of competition and restructuring proposals. A thorough review is provided in Chapter 2.

The electricity industry has been the object of complex, interweaving, and overlapping federal and state regulatory authority ever since New York and Wisconsin initiated state regulation of electric utilities around the turn of the century. A key initial piece of federal legislation was the Public Utility Holding Company Act, passed in 1935, which transformed a complicated industrial structure into the system of state-regulated utilities we know today. In 1977, the Federal Power Commission, empowered by the Federal Power Act of 1935 to regulate the rates and terms of wholesale sales and the use of the transmission network, was reconstituted as the Federal Energy Regulatory Commission, or FERC.

Today's electricity industry is the result of nearly ninety years of government regulation and legislation intended to improve the system or meet some political goal.

In 1978, Congress passed the Public Utilities Regulatory Policies Act (PURPA). PURPA created a new category of electricity generators known as "qualifying facilities" and required public utilities to purchase power from these qualifying facilities at prices no more than the cost of other generating options available to the utility. PURPA opened the door to greatly expanded participation by nonutility generators in electricity markets, and it demonstrated that electricity from nonutility generators could successfully be integrated with a utility's own system supply. (For more about qualifying facilities and nonutility generators, see the box in Chapter 2 on pages 32–33.)

More recently, the enactment of the Energy Policy Act of 1992 created a category of electricity generators exempt from the Public Utility Holding Company Act and has provided an important stimulus for the current regulatory reform efforts.

Implementing Competition

The adage, "Easier said than done," aptly characterizes efforts to reform regulation of the electricity industry and to promote expanded competition. Three factors peculiar to the electricity industry are largely responsible for the difficulties.

1. *Unlike other network industries, such as natural gas and telecommunications, the flows of electricity across the network of interconnected power lines cannot be directed.* This means that changes in the amount of electricity generated by a

utility will increase the electricity flows through the lines of neighboring utilities and that additions to the transmission system of one utility will decrease power flows at neighboring utilities.

2. *Electricity is a unique commodity in that it must be produced largely upon demand so that blackouts don't occur.* Moreover, with current technology, electricity cannot be economically stored. Consequently, the electricity delivery system requires activities that "balance" the electricity load (in other words, guarantee that supply and demand are always equal) at each and every moment in time.

3. *Development of efficient markets for competitive power will require regulatory reforms and coordination by both state and federal regulatory agencies.* Since the electricity industry is subject to overlapping state and federal regulation, jurisdictional disputes—motivated by different political interests and legislative mandates—are likely to slow and complicate the reform process.

In Chapter 3, we discuss the important question of whether to permit expanded wholesale competition or whether to implement the much broader concept of retail competition (which would open competition for retail sales directly to factories, energy marketers, and perhaps small businesses and individual homes). Under expanded wholesale competition, unregulated electricity generators would be permitted to sell power directly to regulated distribution companies and other wholesale purchasers; these distribution companies would appear very much like the electric utilities that now provide power, but they might not own generating facilities as utilities do today. Under the broader concept of retail competition, power producers could sell electricity directly to final customers.

Some analysts argue that the benefits of efficient, low-cost, and high-quality electricity service can be attained under wholesale competition, while others argue that the benefits of competition will be forthcoming only if full retail competition is permitted. Moreover, there is extensive controversy within the industry as to how much this competition would have to be managed to maintain load balances efficiently. The lines of the debate are largely set by the opposing views on these questions.

Transmission Pricing

Between the electricity generator and the distribution company or customer stretches the system of high-voltage transmission lines, often termed simply the *grid*. Currently, use of the transmission grid and pricing of transmission services are regulated by FERC, while distribution to residential, commercial, or industrial customers is regulated by state agencies. While many analysts argue that power generation is no longer a natural monopoly, the transmission and distribution segments of the electricity industry seem likely to remain natural monopolies for the foreseeable future. If that is true, there will be continued pressure to regulate both access to the grid and the pricing of transmission and distribution services so that the benefits of competition in wholesale or retail power markets may be realized.

Issues regarding access to the grid, the pricing of transmission services, and other aspects of transmission regulation are the topic of Chapter 4. We discuss transmission by analyzing pivotal questions: whether to regulate transmission at all, what form the regulation should take, how access and pricing will be determined, and what physical components of the transmission system should be regulated.

Breaking Up the Integrated Utility

As we have noted, the phrase "restructuring of the electricity industry" refers to a range of changes in the ownership or business practices of electric utilities in order to separate the generation, transmission, and distribution functions. Several methods for separating these functions have been proposed, including the following: coupling open-access rules with the virtual unbundling of generation and transmission; creating separate generation and transmission subsidiaries; and divesting transmission companies of their generation businesses. It is not clear yet whether restructuring will be a natural outgrowth of competition, nor is it clear whether policymakers should exercise some control over the process. It is important to understand whether restructuring will enhance or degrade the efficiency of the electricity system in the future, as well as what type of restructuring might be optimal.

Chapter 5 examines the "disintegration" of current "vertically integrated" electric utilities that might be brought about by restructuring. The pros and cons of the vertically integrated utility structure are presented, and we pay particular attention to the anticompetitive influences that vertically integrated utilities might exert on the emergence of deregulated, competitive electricity markets.

Stranded Costs

Consider the case where, because of prior investments in power plants or existing contractual obligations, a utility can break even only by selling power at a regulated price of eight cents per kilo-

watt-hour. If, under newly enacted competition, the price of electricity falls to six cents per kilowatt-hour, the utility will have
costs it can't cover. These are called *stranded costs*. Stranded costs
are just about the most contentious issue in the debate over
expanding competition in the electricity industry.

Many industry analysts believe that competition-induced
declines in price will render unprofitable many generating facilities and existing contracts with third-party generators. If so, advocates of expanded competition are left with the sticky problem of
determining who should suffer the loss—utility investors, nonutility investors, potential new suppliers, current and future customers, or some combination of the above.

The potential size of the costs stranded through competition—perhaps as much as $200 billion—has made the issue of
who will pay a highly contentious one, to say the least. In Chapter 6, we take up the problem of stranded costs and examine the
policy options and the economic factors that should be taken into
account in determining who should pay and how those payments
should be made.

Effects on the Environment

No matter how the electricity industry is restructured, the airborne pollutants from the industry certainly will still be significant. Since the harm caused by air pollution depends not only on
the quantity of emissions but also when, where, and how the
emissions are discharged, predicting the effects that expanded
competition will have on environmental quality is an extremely
difficult task. The relationships that link industry restructuring to
environmental quality are discussed in Chapter 7.

Increased competition in the electricity industry has the
potential to affect the environment in both negative and positive
directions. These effects, which must be balanced to come to a
realistic assessment of the environmental impact of restructuring,
include the following:

- Lower electricity prices will lead to higher electricity consumption and perhaps to higher levels of emissions; this
  suggests the possibility of further environmental degradation. On the other hand, the increased use of electricity
  by commercial and industrial customers may result in a
  reduction in the use of more-polluting forms of energy.
- By increasing access to power markets, restructuring is
  likely to encourage new suppliers to enter into electricity
  generation. Over time, this entry might alter the age pro-
  file of generating plants, perhaps favoring newer, less-

polluting facilities. In this instance, restructuring might serve to improve environmental quality.

- Some recent innovations in pollution regulation (such as the use of tradable emission allowances to regulate sulfur dioxide emissions) may become more effective and perhaps even lead to other innovations that exploit market incentives to limit harm to health and the environment.

- Finally, the future viability of demand-side management programs, which are designed to shift electricity demand to different times of the day or to reduce demand entirely, must be assessed, as must the future role for renewable and less polluting technologies.

All of these relationships will have some bearing on environmental quality; unfortunately, given our current understanding of the relationships, we can do little more than speak about them in a qualitative fashion.

The Goal

Over the next few years, the issues outlined here—how and whether to implement competition, to regulate transmission, to restructure the electricity industry, and to compensate for stranded costs—will stand out as crucial subjects for academic research, industry development, and public policy. For such research, development, and policy to be effective, however, the fundamental factors affecting these issues should be made as clear as possible to the people in government, commerce, the research community, and the public at large, who must both make possible and live with the results of the forthcoming restructuring of the electricity industry. We hope this book contributes to that goal.

Chapter 2

The Electricity Industry and Its Regulatory History

A basic grounding in the functions and structure of the electricity industry and in how it is regulated is helpful to understand much of the debate surrounding efforts to bring competition to the industry. The growing intensity of this debate obscures an important fact: the issues under consideration have been argued about for a long time. To some degree, these issues have played a role in the development of the electricity industry since its inception. To put today's controversies into perspective, in this chapter we present a brief overview of the functions of the electricity industry, the composition of the industry, and the history of public policy regarding the electricity industry, including a description of current regulation.

Functions of the Electricity Industry

In the process of supplying electricity to its customers, the electricity industry performs three primary functions: generation, transmission, and distribution. Some industry experts, in anticipation of retail competition, identify a fourth function by dividing the distribution function into two parts: physical delivery of electricity and retail sales of electricity. While many electric utilities engage in all four of these activities, a large number specialize in one or two. In addition, a growing share of the generation function is being undertaken by *nonutility generators*—generators either unaffiliated with any retail utility or unaffiliated with the utility that ultimately distributes the power. In this section, we offer a brief description of each of the four functions of the electricity industry. (See also figure on page 17.)

Generation

Generation is the process used to create electricity. In most types of generation, some form of energy is expended to drive a turbine, which in turn drives a generator, which in turn produces electric current. The form of energy that is used is a common way of classifying generation facilities.

In 1994, U.S. electric utilities produced 2.9 trillion kilowatt-hours (kWh) of electricity. More than two-thirds of this amount was generated using steam from a boiler, which burns coal, natural gas, or oil to drive a turbine, which in turn drives the electricity generator. A little less than a quarter of the electricity produced came from nuclear power plants, which use nuclear fission of uranium to make the steam that drives a turbine to power the generator. A little more than one percent of the electricity produced in 1994 was generated by gas turbines and small internal combustion engines; a gas turbine uses the combustion of natural gas or distillate oil to turn the turbine that drives the electricity generator. The remaining electricity was generated by hydroelectric plants (which use the flow of water to spin turbines connected to generators), as well as several different renewable technologies, including geothermal, solar, wind, and biomass. (See figure on page 19.)

The economics of electricity generation are changing substantially with the introduction of new technologies, such as high-efficiency gas turbines and combined-cycle gas turbines. Traditional coal-fired, steam-powered generating units need a capacity of 300 to 600 megawatts in order to exploit the economies of scale in generation. These economies of scale occur because, when the larger generators were built, they had both lower average construction costs per megawatt of capacity and lower operating costs per megawatt-hour of generation than did smaller generators. (For a discussion of economies of scale, see box in Chapter 5 on page 86.) However, new combined-cycle gas-turbine plants with capacities as small as a hundred megawatts have lower combined capital and operating costs per megawatt-hour of electricity produced than new coal-fired generators, and they have comparable reliability in generating performance. New gas turbines, which

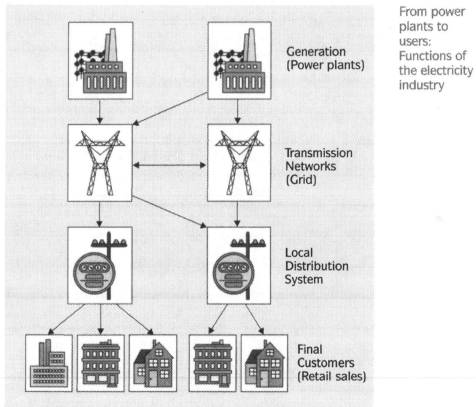

Generation
(Power plants)

From power plants to users: Functions of the electricity industry

Transmission Networks (Grid)

Local Distribution System

Final Customers (Retail sales)

Note: Arrows indicate the variety of paths electricity can take between generators and customers.

were derived from jet engine technology with capacities as small as twenty to fifty megawatts, also may have lower average costs than coal plants, depending on the relative prices of natural gas and coal. These new technologies are making significant inroads into generation activities; in 1994, they contributed to a 35 percent reduction in the average size (total kilowatts of capacity) of new fossil-fuel-fired generating plants relative to the average size of the existing stock of utility generators. This reduction in the minimum efficient size of generating plant means that electricity generation is becoming more amenable to competition.

Transmission

Transmission is the process of conducting the flow of electricity at high voltages from the points of generation to the locations of groups of electricity users, such as residential neighborhoods, industrial parks, or commercial centers. (Large industrial users may also take electricity directly at high voltage from transmission networks.) The transmission system consists of transmission lines, substations with voltage transformers, circuit breakers, and other equipment necessary to transmit power. In addition to delivering electricity, transmission networks or *grids*, as they are sometimes called, connect utilities and facilitate electricity sales among them. Electricity transmission involves substantial fixed costs associated with obtaining rights-of-way and constructing transmission lines and, therefore, the transmission of electricity in any given geographical area is a natural monopoly. (For more on monopolies, see the box in Chapter 4 on page 65.)

> *The electricity industry performs three primary functions—generation, transmission, and distribution. Many experts divide distribution into local delivery and retail sales. Utilities that perform all four functions serve most customers in the United States.*

Distribution and Retail Sale

Distribution of electricity is the process of transforming high-voltage electricity to lower voltages and then physically delivering it to households, industrial facilities, commercial establishments, government offices, and other electricity users. Like transmission, distribution of electricity is usually considered a natural monopoly, and utilities that engage in distribution typically are granted an exclusive service territory by state regulators.

Retail sale of electricity is the process of marketing of electricity to the ultimate customers. Retail sale and distribution of electricity traditionally have been provided as joint products; virtually all retail electricity customers purchase electricity from the utility that delivers it to their premises. If, for coordination purposes or other reasons, it is cheaper for one entity to be both the electricity distributor and the electricity merchant, the current arrangement may prove to be the most efficient. However, while physical distribution of electricity involves large fixed costs for capital equipment, retail sale of electricity does not and therefore may be amenable to competition. For this reason, some analysts have suggested that the retail sale of electricity could be separated from distribution.

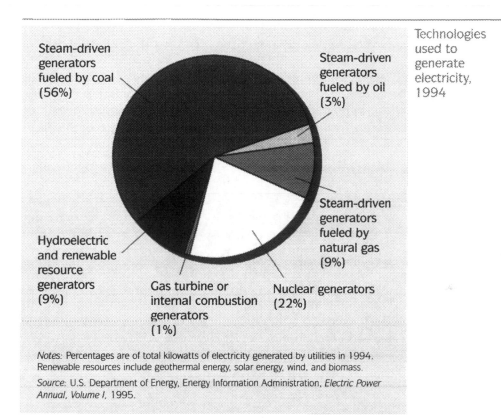

Steam-driven generators fueled by coal (56%)

Steam-driven generators fueled by oil (3%)

Technologies used to generate electricity, 1994

Steam-driven generators fueled by natural gas (9%)

Hydroelectric and renewable resource generators (9%)

Gas turbine or internal combustion generators (1%)

Nuclear generators (22%)

Notes: Percentages are of total kilowatts of electricity generated by utilities in 1994. Renewable resources include geothermal energy, solar energy, wind, and biomass.

Source: U.S. Department of Energy, Energy Information Administration, *Electric Power Annual, Volume I,* 1995.

Composition of the Electricity Industry

The electricity industry is made up of several different types of companies, including both utilities and nonutility generators. Currently, there are about 3,200 utilities in the United States; these utilities may be owned by investors, by state or local governments, by rural electric cooperatives, or by the federal government. (The figure on page 20 shows the generating capacity and the share of sales to final customers for each type of utility.)

Most of the electricity purchased by final users in this country comes from investor-owned utilities that are *integrated*—that engage in all aspects of electricity supply, from generation and transmission to distribution and retail sale. Most of the public and cooperative utilities, however, specialize in distribution and retail sale, while most federally owned utilities specialize in generation. This specialization is apparent in the figure on page 20 as the difference between the share of generating capacity and the share of final sales to customers for each type of utility.

This specialization among the types of utilities has led to a large and growing wholesale market for electricity in the United

States. In 1994, over 1.4 trillion kWh of electricity were exchanged in wholesale markets. The bulk of this trading activity represents purchases under long-term contracts by retail utilities from generating utilities. In addition, integrated utilities and distribution utilities may buy power under long-term contracts from other integrated utilities with excess generating capacity. Much of the remaining trading activity represents short-term sales of excess generation between integrated utilities. In these short-term markets, electricity often sells for prices that are not much higher than the cost of the fuel required to generate it.

Between 1990 and 1994, the quantity of electricity traded at the wholesale level grew by 29 percent; during the same time, sales of electricity to final customers grew by only 8 percent. While most of the increase in wholesale trading was due to sales among utilities, the amount of electricity that utilities purchased from nonutility generators grew by an estimated 80 percent. Throughout the 1980s and early 1990s, many integrated utilities—responding to changes in both federal and state regulation—chose to meet anticipated growth in demand by buying the power they needed from nonutility generators rather than by building their own new generating capacity.

Different types of utilities and their relative generating capacities and share of final sales, 1994

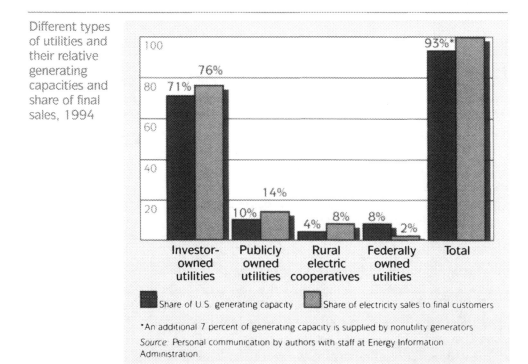

Share of U.S. generating capacity   Share of electricity sales to final customers

*An additional 7 percent of generating capacity is supplied by nonutility generators

Source: Personal communication by authors with staff at Energy Information Administration

An important aspect of this switch to buying rather than building is that almost all of the electricity is purchased under long-term contracts. Many of these contracts were awarded as the result of a competitive bidding process, in which a utility solicited bids from many potential suppliers and then selected the supplier or generator offering the best package of low-cost and high-quality power. The winning generator then would construct a generating plant that usually is dedicated exclusively to supplying electricity to one utility for the next ten or twenty years.

In a few states where this type of competitive bidding has become standard practice, long-term wholesale generation markets to supply new demand are becoming very competitive. Most of the electricity delivered to consumers, however, still is generated by the distributing utility and, therefore, not priced in competitive markets at the wholesale level.

A Century of Electricity Policy

The current structure of the electricity industry represents a substantial evolution from the industry's earliest days. This evolution has been shaped in large part by a ninety-year series of federal and state laws and regulations that govern much of the way electric utilities do business. The history of the electricity industry and of the important legislative and regulatory initiatives that shaped its history are summarized next. (See also the timeline on page 22.) As we shall show, much of the debate that surrounded earlier policy initiatives is relevant for the current policy debate on restructuring.

Beginnings: The Public Utility Holding Company Act and Federal Power Act

The U.S. electricity industry began as an unregulated private enterprise. Wall Street investors financed Thomas Edison's—and the nation's—first power plant, the Pearl Street Station in lower Manhattan, which began to generate electricity in 1882. Private power systems multiplied rapidly thereafter, accounting for 2,800 of the 3,620 systems in operation by 1902. While many municipal governments oversaw the operations of utilities within their borders, the industry remained free of state government oversight until 1907. That year, both New York and Wisconsin established the institutions and rules by which utilities would conduct business in those states—institutions and rules that became models for the widespread system of state public utility commissions. In addition, the privately owned electric utilities actively *sought* gov-

Milestones in
the history of
the electricity
industry

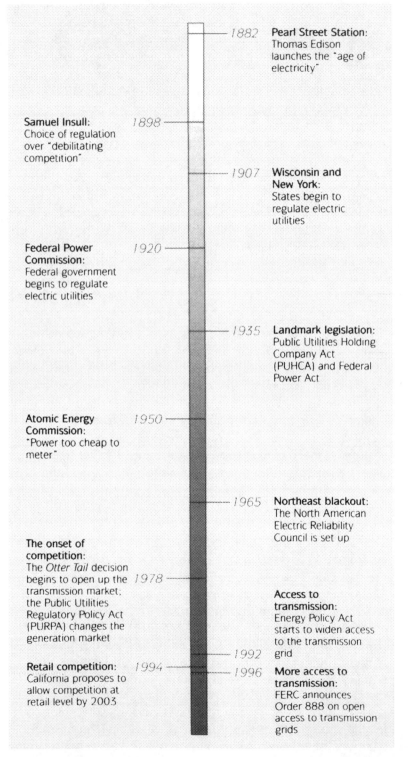

1882 **Pearl Street Station:** Thomas Edison launches the "age of electricity"

Samuel Insull: Choice of regulation over "debilitating competition" — *1898*

1907 **Wisconsin and New York:** States begin to regulate electric utilities

Federal Power Commission: Federal government begins to regulate electric utilities — *1920*

1935 **Landmark legislation:** Public Utilities Holding Company Act (PUHCA) and Federal Power Act

Atomic Energy Commission: "Power too cheap to meter" — *1950*

1965 **Northeast blackout:** The North American Electric Reliability Council is set up

The onset of competition: The *Otter Tail* decision begins to open up the transmission market; the Public Utilities Regulatory Policy Act (PURPA) changes the generation market — *1978*

Access to transmission: Energy Policy Act starts to widen access to the transmission grid

1992

Retail competition: California proposes to allow competition at retail level by 2003 — *1994*

1996 **More access to transmission:** FERC announces Order 888 on open access to transmission grids

ernment protection from what Samuel Insull, one of the industry's founders, called "debilitating competition" and to counteract growing municipal ownership.

The individual owners of the early power plants soon made way for investor-owned companies, which in turn evolved into larger enterprises. During the early part of the twentieth century, holding companies consolidated the industry and substantially reduced the number of firms engaged in the electric services industry nationwide. By the time of the Great Depression, however, these holding companies had proven unstable and their financial structures disastrous. The Depression-era collapse of utility stocks spurred investigations by both the Securities and Exchange Commission (SEC) and the Federal Trade Commission; the results of these investigations crystallized public mistrust of the electricity industry. The SEC publicly charged that the holding companies had been guilty of "...stock watering and capital inflation, manipulation of subsidies, and improper accounting practices." The general counsel of the Federal Trade Commission went further, claiming that "[w]ords such as fraud, deceit, misrepresentation, dishonesty, breach of trust and oppression are the only suitable terms to apply."

These investigations led to the first significant federal effort to restructure the electricity industry. Initially proposed by President Franklin Roosevelt's New Dealers and aggressively implemented by the SEC, the Public Utility Holding Company Act (PUHCA) in 1935 transformed the multistate, complex holding companies into simple corporate structures subject principally to regulation by state authorities. PUHCA granted the SEC broad power to confine acquisition of assets to geographically defined areas and to functions related to utility operations, to prohibit loans from an operating utility to a parent holding company, and otherwise to regulate contracts among affiliates of a holding company. After laboring over the implementation of PUHCA for more than a decade, the SEC created the vertical electric utility structure that dominates today, in which a single company generates electricity, transmits it to cities and towns, and distributes it to factories, businesses, and homes.

While satisfied that PUHCA would effectively control the utilities' corporate structures, President Roosevelt also believed it essential to ensure control of utilities' commerce. The Federal Water Power Act, enacted by Congress in 1920 to regulate the sale of surplus power from federal irrigation and water control projects, created the Federal Power Commission (FPC) and was amended in 1935 to include all wholesale sales of electricity. The amended statute—renamed the Federal Power Act—endowed the FPC with broad authority to oversee the interstate transmission of electric energy and to establish rates to be charged for wholesale sales of power. The FPC also was given limited authority to order interconnection among utilities to enable delivery of electricity from one service area to another. In addition, the FPC was authorized to forbid overlapping boards of directors across power companies and to approve the sale of facilities and mergers.

No subsequent amendments or enactments of other laws have substantially reduced the powers of the FPC to regulate the electric utility industry. In 1977, with the creation of the U.S. Department of Energy, the FPC was reconstituted as the present-day Federal Energy Regulatory Commission (FERC), which is composed of five commissioners appointed by the President and confirmed by the Senate.

The Federal Role in Providing Electricity

While Congress resisted New Deal proposals to nationalize the electricity industry, nonetheless President Roosevelt obtained enactment of statutes that created a federal electric utility sector that would operate alongside investor-owned and municipal utilities. In 1933, after long and divisive debate about the proper role of government, Congress created the Tennessee Valley Authority (TVA), which eventually became the nation's single largest utility. Although its establishment was justified on grounds of flood control, job creation, and rural development, TVA's central mission from the outset was to provide electricity to towns and rural cooperatives. Congress provided grants, subsidized low-interest loans, and granted tax exemptions to TVA power systems. At the same time, Congress gave municipalities and rural cooperatives preferential access (which was not granted to private utilities) to the subsidized low-price electricity produced at TVA and other federal facilities, as well as additional tax exemptions.

In 1937, Congress created the Bonneville Power Administration (BPA) in the Pacific Northwest. BPA was not a combined development agency and public utility like TVA, but rather a marketer of electric power generated at federally constructed dams on the Columbia River system. Like TVA, however, BPA could borrow from the U.S. Treasury at below-market interest rates, and municipal utilities and rural cooperatives, whose power systems also were federally subsidized, had preferential access to its power. BPA was followed, in later years, by similar power marketing administrations serving the West, the Southwest, the Southeast, and Alaska.

Postwar Complacency and Stability

By 1950, federally generated power was reaching one in four American households. Privately owned utilities were fully under the control of state public utility commissions, the SEC, and the FPC. These regulators were gratified by the stability provided by what they considered America's unique "pluralistic structure" of public and private power. Driven by steady improvements in technology, average electricity rates during the 1950s fell from 3 cents/kWh to 2.5 cents/kWh for residential customers, and demand for electricity grew at twice the annual rate of the national economy. New technological opportunities surfaced for the industry with President Dwight D. Eisenhower's Atoms for Peace program and the subsequent construction, in 1957, of the nation's first commercial nuclear-power generating plant at Shippingport, Pennsylvania. As electricity demand continued to grow in the postwar era, the 1950s concluded with the development of regional system coordination and power pools. These organizations provided utilities with an easy means of getting electricity to those with sudden needs from those who could generate additional power to meet those needs.

In the industry's early years, privately owned electric utilities actively sought government protection. Through the 1950s, the regulatory structure seemed to satisfy both the industry and government regulators.

The 1950s were the calmest decade the electricity industry was to enjoy. The regulatory structure satisfied the industry, because the industry was allowed to go about its business with few other impediments. Utilities were required only to keep adding to generation and transmission capacity. State regulators facilitated siting of new plants and transmission lines and seldom, if ever, second-guessed their own or the utilities' decisions. Congress also was satisfied with the utility structure that resulted from the New Deal, refusing Eisenhower's insistent proposals to privatize TVA and the power marketing associations. The inherent inflexibility of the system was yet to be felt.

In 1964, the FPC's National Power Survey projected growth in electricity demand that would require increasing generation capacity from 200 gigawatts in 1964 to 525 gigawatts in 1980 (a gigawatt is a billion watts). The survey projected a dramatic reduction in the capacity of industrial firms to generate their own electricity, a practice that during World War II had provided for half of this sector's power needs. The survey also predicted that seventy gigawatts of inexpensive nuclear technology capacity would be built by 1980. Total generation costs by 1980 were predicted to be on the order of 0.5 cents/kWh, including capital investment, fuel, and operating expenses.

In its 1964 survey and report, however, the FPC warned that far better coordination would be needed to manage this much larger, complex, and interconnected system. The warning was both prophetic and late. On November 19, 1965, the United States suffered the great Northeast Blackout, affecting thirty million people from Canada to New Jersey. In the aftermath, the news media characterized the industry as incompetent and moribund.

The utility industry responded by voluntarily constituting what came to be known as the North American Electric Reliability Council (NERC). NERC was (and is) made up of ten distinct regional councils charged with keeping electricity service reliable in the three highly synchronized electric transmission grids that span the United States, Canada, and parts of Mexico. (For a list of the regions and grids, see box on page 27.) Each regional council was responsible for system reliability in its geographic area of oversight. Utilities were forced to cede to the councils a degree of operational freedom both as distinct generators and transmitters of power and as members of power pools. These concessions, however, produced a system that achieved unprecedented levels of efficiency in the generation and dispatch of power through rigorous coordination of regional capacity and distribution.

The Energy Crisis: Industry and Regulators Respond

The 1970s were the decade of crisis for the electricity industry. The industry and its regulators had come to expect predictable behavior from each of the key agents in the electric fuel cycle and were unprepared for the social, political, and economic upheavals touched off by the Arab oil embargo of 1973. The tripling of oil prices precipitated the need for rate increases, since oil fueled many power plants. In 1974, the year following the embargo, utilities filed an unprecedented 212 requests for rate increases totaling $4.5 billion; state regulators approved increases amounting to only about half that. Applications for rate increases continued to be filed in record numbers throughout the 1970s and well into the following decade; in most instances, the increases allowed by public utility commissions continued to fall short of the utilities' requests. Nevertheless, rate increases sharply reduced the rate of growth of demand for electricity.

When this decline in demand growth was combined with increased use by state regulators of prudence reviews (where regulators could refuse to allow utility investors to recover the cost of plants), cancellation or postponement of plant construction became inevitable. Between 1972 and 1984, a total of 113 nuclear plant projects were scrapped, as well as sixty-seven coal-fired units. Widespread cancellation of nuclear projects also was attributable to cost increases resulting from increased public opposition to nuclear power and more stringent regulatory oversight in the wake of the nuclear power plant accident at Pennsylvania's Three Mile Island in 1979.

North American Electric Reliability Council

The North American Electric Reliability Council is made up of ten regional councils charged with keeping electricity service reliable in three highly synchronized electricity transmission grids that span the United States, Canada, and parts of Mexico.

The ten regional councils are:
- Alaska Systems Coordinating Council (ASCC)
- East Central Area Reliability Coordination Agreement (ECAR)
- Electric Reliability Council of Texas (ERCOT)
- Mid-America Inter-connected Network (MAIN)
- Mid-Atlantic Area Council (MAAC)
- Mid-continent Area Power Pool (MAPP)
- Northeast Power Coordinating Council (NPCC)
- Southeastern Electric Reliability Council (SERC)
- Southwest Power Pool (SPP)
- Western Systems Coordinating Council (WSCC)

The three transmission grids are:
- Eastern Intertie
- Western Intertie
- Texas

The crisis decade brought forth many proposals from industry and academia for restructuring the electric utility industry. In 1975, a group of New York utilities proposed the creation of a "pure" power-generation company that would operate separately from transmission and distribution enterprises. The New York State public utility commission rejected the proposal, in large part because the new entity would have been entirely under federal control. In the same year, Michigan public officials actively contemplated the complete takeover of the state's electric utilities. In

1978, Matthew Cohen and William Berry of the Massachusetts Institute of Technology (MIT) proposed several prototypes of entirely deregulated power generation, with regional power brokers to operate and maintain the bulk power transmission system and to serve other market functions. These and later proposals offered by Paul Joskow and Richard Schmalensee, also of MIT, consistently held to the premise that while power generation may no longer be considered a natural monopoly, transmission and distribution functions would continue to be and would therefore require regulation.

None of the restructuring proposals from the 1970s found favor with Congress or with FERC commissioners. While major moves to interject competition into the electricity business were yet to come, the courts and antitrust authorities signaled their interest in the electricity industry. In 1979, the Supreme Court handed down a decision in *Otter Tail Power, Inc. v. United States* that forced investor-owned utilities to "wheel" power to a municipally owned system. (For a description of wheeling, see box on page 29.) A broad reading of the *Otter Tail* decision suggests that industry arrangements for transmitting electricity are within the reach of the antitrust laws. At the same time, the Antitrust Division of the U.S. Department of Justice sought to break up the vertical structure of the telephone industry to make it more competitive—an early warning of the concerns now prominent in the policy debate about restructuring the electricity industry.

The Public Utilities Regulatory Policies Act (PURPA)

Interest in changing the electricity industry intensified with the election of President Jimmy Carter in November 1976. Many of his advisers sought means to extend substantially federal authority over the electric utility sector. Their results were codified in the National Energy Act of 1978, a compendium of laws aimed at restructuring the entire U.S. energy sector, with several features directed specifically at electric utilities.

By far, the Carter administration's most significant contribution to restructuring the electricity industry, however, was the Public Utilities Regulatory Policies Act (PURPA) in 1978. As initially proposed by Carter, PURPA would have created an entirely new category of electricity generators exempt from PUHCA—a model adopted fourteen years later in the Energy Policy Act of 1992. The original PURPA would have required states to reform retail electricity rates by adopting *marginal cost* pricing—that is, by basing rates on the actual cost of the last units of power generated, which generally came from the most expensive plants to

What is wheeling?

Wheeling, in the electricity industry, refers to the practice of a utility that owns a transmission network taking in and passing along—in other words, delivering—electricity produced by another utility or generator.

Delivering electricity is not quite the same as, say, delivering goods with a truck, because of a peculiar property of electricity, namely, that once electricity enters the transmission system, it all becomes the same—it is impossible to deliver exactly Generator A's electrons to Customer B. (For more about the peculiar properties of electricity and how they affect transmission issues, see Chapter 4.)

The obligation of utilities to wheel power through their transmission facilities is essential to a more competitive electricity industry, because wheeling allows distant generators to compete for customers. The right to have power wheeled was established in the 1977 Supreme Court case, *Otter Tail Power, Inc. v. United States*. In that case, the Court found that Otter Tail could not refuse to wheel federally marketed power to the municipal distribution utility in Elbow Lake, Minnesota.

operate. (For more about marginal cost pricing, see box in Chapter 4 on page 72.) PURPA also would have set standards for interconnections, wheeling, and pooling; banned discounts for large power users; and required utilities to purchase power produced by *cogenerators*—factories that produce electricity as a by-product of their primary manufacturing processes—and from plants using renewable fuels.

In the end, Congress curbed PURPA's proposed sweep. As enacted, PURPA made rate reform voluntary. That is, each state could decide for itself whether to set prices to reflect time-of-day use or to give discounts to users willing to accept interruptions in service. A primary focus of the new law was the creation of *qualifying facilities*, which comprised cogenerators and small power producers that used renewable energy sources. PURPA obligated utilities to connect qualifying facilities to transmission grids and to purchase their power at a price that did not exceed the *avoided cost* of new capacity—that is, how much money the utility would save by not having to build new plants to provide what they could get from the qualifying facilities.

Utilities also were required to supply standby power at nondiscriminatory rates to qualifying-facility cogenerators. To promote further the renewable power advocated in PURPA, Congress separately enacted the Energy Tax Act, providing a 10 percent tax credit to generators using biomass, geothermal, wind, and solar energy. The Power Plant and Industrial Fuel Use Act, enacted concurrently with PURPA, forbade the use of oil and natural gas in new power plants. By eliminating these preferred fuel and technology options for utilities' peak load requirements, it increased demand for electricity from nonutility generators, including qualifying facilities and independent power producers. (For a description of the new categories of generators that evolved in the wake of PURPA, see box on pages 32–33.)

The Northeast Blackout in 1965 and the "energy crisis" of the 1970s inspired proposals to restructure the industry. With the passage of PURPA in 1978 and EPAct in 1992, new players entered the market for generation.

While PURPA fell short of bringing full competition to electricity generation, it did demonstrate that electricity from nonutility generators could be integrated successfully with a utility's own supply system. Qualifying facilities were able to secure funding in financial markets, to bring capacity on line in a timely fashion, and to operate reliably. Utilities, which had often become reluctant to construct new power plants as a result of past disallowances, turned increasingly to nonutility generators as a source of power. In 1986, nonutility generators were contributing 20 percent of new generation capacity; by 1994, their share had grown to over 60 percent of new capacity additions. Of the 7 percent of total U.S. generating capability supplied by nonutility generators in 1994, qualifying-facility cogenerators and small power producers accounted for roughly 80 percent; independent power producers and non–qualifying-facility cogenerators made up the remaining 20 percent.

Regulation-Induced Competition and the Energy Policy Act

The rules governing utility purchases from qualifying facilities tended to limit any benefits, such as lower generation costs and lower electricity rates, from competition among these nonutility generators. The statutory requirement that all power from qualifying facilities be purchased at a price that did not exceed utilities' avoided cost was interpreted by FERC as a requirement to ignore whether the benefits of purchasing power from qualifying facilities exceeded the costs. Many states went one step further by requiring utilities to purchase *all* power offered by qualifying facilities at a price less than or equal to the utility's avoided cost.

Qualifying facilities proliferated in the five years following passage of PURPA because their owners could negotiate contracts with utilities on terms made lucrative by the elastic definition of "avoided cost." The avoided cost for each utility was set by or in consultation with the state regulator. In some states, a utility's avoided cost of acquiring new generation capacity was based on the cost of, in some cases, a new nuclear power plant or, in other cases, a re-powered coal-fired plant. In still other cases, it was equated to the highest-cost, peak power plant. Imperfect from the start, calculations of avoided costs became increasingly untenable as the high-priced energy of the early 1980s gave way to reduced fuel prices and cheaper generating technologies.

The need to determine a reference price for PURPA power and to establish a benchmark for what the market might consider to be the avoided cost of new generation capacity led to the establishment of competitive bidding processes. Between 1984 and 1991, thirty-six states adopted or considered competitive procurement procedures to acquire new capacity. Sixty-seven requests for proposals were issued in the period. In the forty-two cases in which the solicitation process was carried to completion, agreement was reached on 315 projects representing nearly twelve gigawatts of generation capacity. These competitive bidding processes, as well as contractual terms negotiated between utilities and qualifying facilities or independent power producers, varied widely in their degree of adherence to market principles. In many cases, state regulators, reluctant to permit the market alone to discipline the bidding process, established complex rules constraining the flexibility of the parties involved to obtain the best available generation technology at lowest cost.

The move toward restructuring and increased competition in generation begun by PURPA was accelerated considerably by the Energy Policy Act (EPAct) in 1992. Under EPAct, Congress required that FERC force transmission-owning utilities to deliver power from generators to other utilities and electricity wholesalers at reasonable, nondiscriminatory, cost-based rates. To carry out this mandate, in April 1996 FERC issued Order 888, which specifies the conditions under which all utilities must provide such access to the U.S. transmission system.

EPAct also allowed the formation of a new class of entities called *exempt wholesale generators*, or EWGs. Unlike the qualifying facilities created by PURPA, these EWGs are exempt both from requirements to use particular fuels or technology and from requirements concerning their corporate structure under PUHCA. Unlike PURPA, EPAct did not require utilities to buy power from EWGs at avoided cost or, for that matter, at any other prescribed level. Thus, EWGs are expected to compete with one another and

New players evolving after PURPA and EPAct

Since the Public Utilities Regulatory Policies Act (PURPA) was adopted in 1978, traditional utilities have been joined by several new players in the generation business.

Nonutility generators (NUGs) are the entire group of power generators that are *not* affiliated with the utility that transmits, distributes, and sells to final customers in a particular area. In 1994, nonutility generators accounted for about 7 percent of U.S. generating capacity.

The distinction between utilities and nonutility generators can be confusing. For one thing, nonutility generators may be owned by electric utilities (and many of them are); however, with the exception of qualifying facilities that are partially owned by utilities, those nonutility generators may *not* sell power within the service territory of the parent company. For another, to some extent, every firm that sells electricity is an electric utility under the Federal Power Act and, consequently, subject to oversight by the Federal Energy Regulatory Commission (FERC). It is easiest to think of a nonutility generator as a company that does not sell to final customers in its function as a generator.

Qualifying facilities (QFs) are nonutility generators that meet certain criteria spelled out in PURPA. Qualifying facilities include *cogenerators*, firms that produce electricity and other forms of useful thermal energy as the result of industrial processes. Qualifying facilities also include small power producers that use a renewable resource (such as water, solar energy, or biomass) to generate electricity.

PURPA required utilities to purchase the power produced by qualifying facilities at a price at or below the utilities' *avoided cost*—what it

with the purchasing utility's own generators for opportunities to sell their electricity to retail utilities and other wholesale buyers.

The Important Role of State Utility Regulators

As of early 1996, the Federal Energy Regulatory Commission is the primary federal regulator of electricity policy. FERC regulates rates for wholesale power sales, regulates rates for electricity transmission, and has the authority to approve or disapprove utility mergers. However, much (and perhaps most) of the important regulatory impetus and initiatives to bring competition to the electricity industry is taking place at the state level. Since the onset of state oversight by New York and Wisconsin at the turn of the century, the rates paid by most electricity customers have been set by state public utility commissions (PUCs). While state PUCs generally do

would have cost the utility otherwise to generate or purchase that additional electricity. PURPA left it to individual states to set the avoided cost. Some states, such as California, set the avoided cost quite high in the beginning, thus encouraging cogeneration and generation with renewable resources.

Independent power producers (IPPs) are nonutility generators that are not qualifying facilities. After PURPA took effect, qualifying facilities sprang up and began selling to utilities; at the same time, several state utility regulators started requiring competition in the purchase of generation, and FERC began to encourage competitive generation among nonqualifying facilities by allowing those generators to charge market-determined rates for their power. This led to the development of independent power producers, entities that were not qualifying facilities but that could compete in selling generation capacity. IPPs could be owned by a utility, as long as the IPP was not selling to the parent utility's final customers.

As competition grew, companies that wished to enter the generation market were somewhat discouraged by the demands of the Public Utility Holding Company Act (PUHCA). A utility or firm in, say, New York, that wanted to set up as an IPP in, say, California, would be subject to PUHCA's daunting requirements for filing and reporting.

Exempt wholesale generators (EWGs) were established under the Energy Policy Act of 1992 to address this problem. EWGs were exempt from PUHCA, but *not* exempt from other regulatory oversight, including that of FERC. Any facility wishing to be an EWG must obtain permission from FERC, which is done, in part, by convincing FERC that the EWG subsidiary does not share cost accounts with the parent utility.

not regulate the retail rates charged by municipal utilities and rural cooperatives, they do regulate the rates charged by investor-owned utilities, which accounted for more than three-quarters of all retail electricity revenues in 1994. In the same year, the revenues from retail sales accounted for more than 89 percent of the more than $177 billion in total revenues of investor-owned utilities. Thus, a much larger portion of investor-owned utility revenues are overseen by state regulators than by federal regulators.

State PUCs not only set retail electricity prices, but also set most of the rules regarding entry into the generation business, as well as the boundaries that define a utility's exclusive service territory. Moreover, in many states, a utility wishing to construct a new generating facility or new transmission line must obtain a certificate of public convenience and necessity from the PUC. In addition, other state regulatory agencies, including environmental agencies, may have an oversight role with regard to siting new generation and transmission facilities. In some states, PUCs and environmental departments design and enforce many of the pollution policies affecting the electricity industry; these policies include "social costing" policies that direct utilities to plan and use power plants in ways that reflect not just direct generating costs, but also the social costs imposed by the plants' emissions. (For more on social costs and the environmental issues involved in restructuring, see Chapter 7.)

In most states, the method used to set retail electricity rates is called *cost-of-service* or *rate-of-return* regulation. This means that rates are set so that the revenues from retail sales of electricity will cover the full costs of supplying that electricity, including generation, transmission, and distribution costs, plus a fair rate of return on invested capital. A major drawback of this method of setting electricity prices is that it does not provide the regulated utility with any incentive to minimize its costs. If the utility is guaranteed revenues sufficient to cover its reasonably incurred costs and, furthermore, if efforts to reduce those costs result in a commensurate reduction in utility revenues, then utilities have little incentive to reduce costs. The threat of regulatory disallowance of some utility costs deemed by regulators to be imprudent has weakened somewhat the guarantee that regulated revenues will be sufficient to cover all costs; however, this method of controlling utility costs is somewhat arbitrary and costly to administer.

In recognition of this and other weaknesses of cost-of-service regulation, several states, including New York, California, and Maine, are beginning to adopt other methods of regulating retail electricity rates that provide greater incentives for utilities to behave efficiently. These alternative methods, often referred to as *incentive-based* regulations, include the following variations (and are discussed in more detail in Chapter 4):

- *price-cap regulation*, which sets a cap on the retail price for electricity;
- *sliding-scale or shared-savings regulation*, which allows utility investors to keep some portion of the profits from cost-reducing activities while sharing the remainder with ratepayers; and
- *yardstick or benchmark regulation*, in which the regulated firm's rates are based on the costs of other firms in similar markets.

By moving to incentive-based regulatory methods, regulators are providing utilities with reasons to be more efficient, behavior that should serve the utilities well as buyers and as sellers in more competitive electricity markets.

Conclusion

The extent to which some or all of the regulatory apparatus summarized in this chapter should be modified or eliminated will depend in no small measure on how the recent trend toward deregulation and competition plays out. The central issues, as we see them, involve the implementation of competition; transmission pricing; vertical integration among generators, transmission companies, and local distribution grids; recovery of prior investments in power production; and environmental protection policies. In the next chapter, we look at proposals for bringing competition to the electric power industry.

Chapter 3
Implementing Competition
Different Levels, Different Models

A crucial consideration in bringing competition to the electricity industry is
establishing a market framework so that
those who produce electricity can sell it
to those who want to use it. For other
commodities, one might be confident
that the process of setting up markets
would take care of itself, but electricity
has two unusual characteristics that
greatly complicate the development of a
competitive market. The first characteristic is one we have mentioned already:
electric power systems require constant
balancing to ensure that the amount of
power users demand is neither more nor
less than the amount of power being generated. The second characteristic is that
the flow of electricity within the power
system cannot be confined to a predeter-
mined path. These two characteristics,

peculiar to electricity, have to be accommodated by any proposal to implement competition in that market. As a result, such proposals must not only define the type of competition that will be implemented but also provide a framework for maintaining load balances and for pricing transmission once competition has been enabled.

In this chapter, we consider a host of issues related to implementing competition in the electricity industry. First, we review the forces at work that make a more competitive industry look attractive. Then we describe the two major types of competition that could be put into effect—expanded wholesale competition and retail competition—and summarize the arguments favoring each type. The bulk of this chapter, however, concentrates on developing a market structure that accommodates the load-balancing requirements of electricity. We describe two different market structures—known as bilateral contracting and Poolco—that could be adopted, and we look at how each structure would operate under either type of competition. We conclude this chapter with a comparison of the two market structures. (Later, in Chapter 4, we take up issues related to transmission pricing.)

Forces for Competition

In Chapter 2, we discussed how advances in technology are making competition an increasingly attractive alternative to traditional regulation of the electricity industry. A second motivation for change is dissatisfaction with the current way regulation sets electricity prices. This dissatisfaction has several components. Some observers claim that regulated electricity prices are artificially inflated to cover the costs of excess generation capacity in the industry and of power purchased under expensive long-term contracts with qualifying facilities mandated by the Public Utilities Regulatory Policies Act (PURPA). This perception is supported by the fact that prices in wholesale electricity markets are often just above fuel costs and, therefore, substantially lower than retail prices in many parts of the country.

Another component of the dissatisfaction with regulation-determined prices for electricity is the substantial variation in price. To some extent, this variation stems from differences in a wide range of factors, including fuel costs, access to low-cost hydro power, utility social programs, state and local tax laws, regulatory implementation of PURPA across states, and the incentives and ability to control costs in a regulated monopoly setting. In 1993, prices ranged from less than four cents per kilowatt-hour in the Northwest to more than ten cents per kilowatt-hour in New York and parts of New England. In addition, the cost of electricity can vary widely both by source and by customer class (see table below). For instance, the prices charged by publicly owned utilities and rural cooperatives—which are often much lower than the prices charged by investor-owned utilities—reflect current and historical subsidies extended to those utilities by the federal government since the 1930s.

Type of utility	Average	Industrial	Commercial	Residential	Average retail electricity prices by customer class (cents/kilowatt-hour, 1993)
Investor-owned utilities	7.2	5.0	7.9	8.8	
State and locally owned utilities	6.1	4.9	6.8	6.6	
Rural electric cooperatives	7.0	4.6	7.4	7.7	

Source: U.S. Department of Energy, Energy Information Administration, *Electric Sales and Revenue,* 1993.

The variation in prices across customer classes reflects differences both in the quantity of electricity used and in the available alternatives to buying from the local utility. Industrial users generally have a larger demand for electricity than commercial and residential users; thus, as a result of economies of scale in distribution, industrial users are less costly to supply. (For more about economies of scale, see box in Chapter 5 on page 86.) Industrial users also pay lower prices for electricity because they have more options if they do not like the price charged by their local utilities. Industrial customers may choose to generate their own electricity, knowing that under PURPA their local utility may be required to buy any excess power that they generate. They also may relocate their production facilities to a region of the country with lower electricity rates. In some situations, large industrial customers have persuaded local community officials to pursue municipal takeover of the local electric utility; when this is done, the entire town has the ability to find a low-cost wholesale electricity supplier or to exploit their preferential rights to subsidized federal power. In many cases, local utilities have responded to such threatened departures by offering discounted electricity rates to industrial users.

Proponents of competition argue that it should reduce—and maybe eliminate—price variations across regions, customer classes, and utility types. Depending on the type of competition introduced, some buyers (local distribution utilities under expanded wholesale competition and electricity consumers under retail competition) should become able to choose among alternative power suppliers. According to advocates of competition, opening the generation market should lead to more rational investment strategies and should drive the search for more efficient ways to generate electricity, as well as for new ways to increase the value of electricity to consumers.

The Scope of Competition

Currently, two approaches to competition are being considered for the electricity industry. The first, and less far-reaching, is referred to as *expanded wholesale competition*. An extension of the wholesale electricity trading already provided for under PURPA and the Energy Policy Act of 1992 (described in Chapter 2), expanded wholesale competition essentially would open the market so that all generators could sell power to local distribution utilities and other wholesale customers, such as power marketers. More radical, and some would argue more necessary, is *retail competition*, in which generators compete to supply power to the customers, either directly or through independent retail power marketers.

Expanded Wholesale Competition

According to the scenarios for expanded wholesale competition, the wholesale market for electricity would be opened up, and any generator that wished to sell electricity would use the wholesale market. Under expanded wholesale competition, the generation, transmission, and distribution functions of integrated utilities would be unbundled so that utilities that distribute electricity to final customers could purchase transmission services from other utilities and generation services from other utilities or nonutility generators. From the perspective of end users, the world under expanded wholesale competition would not look much different from today's. Customers would still purchase electricity as they do now—from a local, state-regulated utility. (See figure on page 41.)

Expanded wholesale competition would deregulate *only* the generation segment of the electricity industry. When the generation function is deregulated, local distribution companies would obtain all the power they sell to households, businesses, and industry via a wholesale purchase of electricity from one or more generating companies. Sales within this generation market would take place at market-determined prices, but local distributors

Wholesale Competition

Under wholesale competition, the customer must purchase electricity from
the local distribution company, which both sells the power and delivers it.

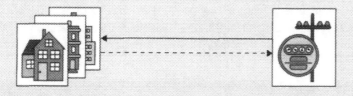

Retail Competition

Under retail competition, the customer can choose to purchase electricity
from one of several marketing companies or directly from a generator.
Electricity is delivered to customers by the local line company.

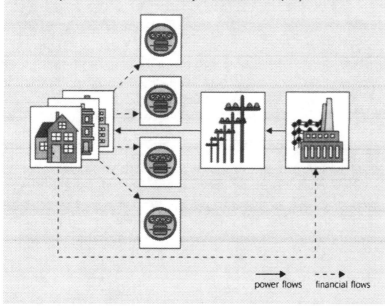

power flows   financial flows

would keep their monopoly franchises for selling electricity to
retail customers in their service territories and would continue
to be regulated by state regulators. Under expanded wholesale
competition, the transmission function, which connects generators to local distributors, would likely continue to be regulated
by the Federal Energy Regulatory Commission, as it is today.

Retail Competition

Retail competition scenarios are considerably more ambitious
than the expanded wholesale competition approach. Under
retail competition, the functional unbundling of transmission
from generation and local distribution services would be

extended to separate the local distribution function from retail sales. From the retail buyer's perspective, the world under retail competition could look quite different than it does today. These end users—perhaps including residential customers—would purchase electricity from the generator or marketer of their choice. There would be several different electricity retailers from which to choose, each offering potentially different packages of services and prices.

Expanded wholesale competition would open the market so that generators could sell power to local distributors and other wholesalers. Under retail competition, generators could supply power to customers directly or via marketers.

Under most scenarios for retail competition, both the upstream generation and the downstream retail sale functions would be deregulated and open for competition. Generating companies would resemble those under wholesale competition, except they would sell power to electricity retailers or directly to customers, instead of to a local distributor with a monopoly franchise for selling power. Power retailers or marketers would buy power from generators and resell it, perhaps bundled with energy management services, to retail customers. The transmission function, bringing electricity from generators to local distributors, would operate and likely be regulated the same as it would under expanded wholesale competition except that power retailers, generators, or customers, not local distributors, would arrange for transmission services. The local distributor or line company would sell its services as a distributor of electricity, presumably under regulation, but need not be the actual retailer of electricity itself.

How Much Competition Is Enough?

Recognizing the pressures for competition, many observers of the electricity industry suggest that it will be difficult to limit competition to wholesale markets alone. Under expanded wholesale competition, electricity consumers may not be sufficiently able to express their willingness-to-pay for different qualities of electricity service and to associate themselves with different electricity suppliers. Given these limitations, expanded wholesale competition may not provide lower retail prices and broader service choices, so then it would make sense to extend competition into the retail markets. Moreover, if product differentiation and marketing are important to electricity users, then extending competition into retail markets becomes even more important.

The degree to which wholesale competition might lead to lower cost and a broader range of services depends on how the local distribution company is regulated. Traditional cost-of-service regulation limits a local utility's incentives to seek aggressively the lowest-cost source of power, because any cost savings from obtaining generation at a lower cost would have to be passed along to customers in the form of lower electricity rates. However, a phenomenon known as *regulatory lag* could counteract this effect. (Regulatory lag refers to the delay between the time when a utility's costs change and the time when regulated prices are adjusted to reflect that change.) Furthermore, the threat that a regulator might disallow imprudent high-cost power purchases also should encourage utilities to consider cost when selecting a generation supplier.

The adoption of incentive-based regulatory methods also could enhance a utility's incentives to reduce its costs under expanded wholesale competition. These methods include price-cap regulation, yardstick regulation, and regulatory shared-savings plans. Price-cap regulation fixes a price ceiling, with adjustments for inflation and expected productivity growth over time. Under yardstick regulation, the utility's price is based on cost data from comparable utilities. In both cases, regulated utilities have a strong incentive to reduce costs, since price is unaffected by costs or profits. If the utility performs better than the price cap or yardstick target, it will profit; if not, it will lose money. Under a regulatory shared-savings plan, utilities pass on a portion of any cost savings to consumers in the form of lower prices and keep the remaining portion for utility stockholders.

Implementation problems may prevent these regulatory reforms from achieving these ideals. For example, the electricity regulator in the United Kingdom, where distribution utilities face price-cap regulation, already has succumbed to political pressure to reduce rates in light of record high utility earnings. Conversely, a regulator may decide to raise prices if the utility is losing money. Recognizing that regulators may give in to these pressures limits a utility's incentive to cut costs. In addition to these pressures, yardstick indices may not accurately reflect cost circumstances that vary across distribution companies. Allowing competition at the retail level may be the best way to enable consumers to seek out the lowest-cost sources of electricity on their own behalf.

Two Proposed Market Structures

Just as there are two types of competition under consideration for the electricity industry, there are also two types of market structures; either market structure could be implemented with either type of competition. One of the market structures, referred to as *bilateral contracting*, envisions a market where most electricity transactions take place under specific supply contracts between two parties: on one side, generators and, on the other side, distributors, marketers, or final customers. The other market structure calls for most short-term power transactions to be coordinated by a centralized spot market, where electricity would be bought as needed and sold as available. This spot market would be run by an institution that has come to be known as a *Poolco*, a term that refers to the coordinator of all suppliers and users of electricity in a common network. For the rest of this chapter, we present each combination of competition and market structure and discuss the drawbacks of each combination, as well as some ways to mitigate these drawbacks.

Bilateral Contracting and Expanded Wholesale Competition

The bilateral contracting structure developed from the current pattern of wholesale electricity transactions. The easiest way to understand how bilateral contracting works under expanded wholesale competition is to know who the key participants are and how electricity gets from the generator to the final user. We present a picture of these participants and transaction paths in the figure on page 45. Even though the picture is simplified, it does capture the essential features of the various proposals for bilateral contracting under expanded wholesale competition (which we refer to as "wholesale bilateral contracting").

The following are the key participants in the model of wholesale bilateral contracting, beginning at each end of the transaction path and working toward the middle. It is important to remember that the separate companies identified here actually refer to separate functions which may be undertaken by the same corporate entity. Arguments for and against breaking up that corporate entity are considered in Chapter 5.

Customers

Customers are the residential, commercial, and industrial users of electricity. They are the end users, and they buy from local distribution companies.

Distribution companies

These local utilities are responsible for purchasing electricity from a generator and then distributing that electricity to customers who do not generate their own power. Distribution companies

sell a bundled product—consisting of power and distribution services—directly to electricity customers. A distributor usually has
a monopoly franchise, issued by the state in which it is located, to
supply electricity to all customers in its service area. In exchange
for this franchise, the distribution company is subject to regulation by the state public utility commission.

Generating companies

At the other end of the transaction, electricity is supplied by any
number of generating companies that compete for the business of
the distribution companies responsible for delivering electricity to
customers. Generating companies arrange to sell electricity
directly to local distributors, or they use the services of brokers or

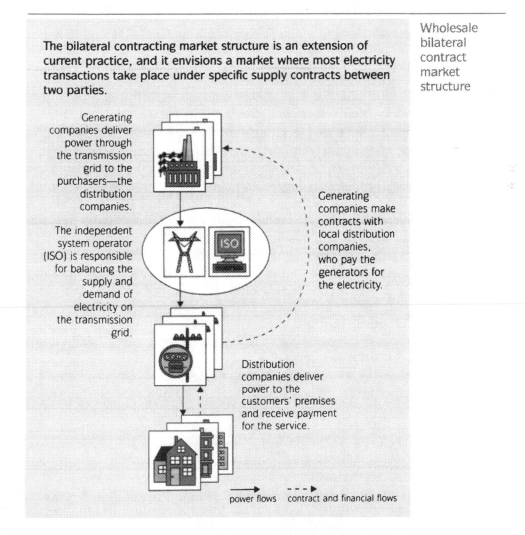

The bilateral contracting market structure is an extension of
current practice, and it envisions a market where most electricity
transactions take place under specific supply contracts between
two parties.

Wholesale bilateral contract market structure

Generating companies deliver power through the transmission grid to the purchasers—the distribution companies.

The independent system operator (ISO) is responsible for balancing the supply and demand of electricity on the transmission grid.

Generating companies make contracts with local distribution companies, who pay the generators for the electricity.

Distribution companies deliver power to the customers' premises and receive payment for the service.

power flows   contract and financial flows

marketers who facilitate transactions with distributors. Some generating companies sell generation services directly to an entity known as the Independent System Operator, which we describe below.

Grid company

The grid company owns and maintains the transmission grid that connects the generating companies and the distribution companies. (The transmission grid is to the electricity industry what the interstate highway system is to the transportation industry.) Under current discussions of wholesale competition, the transmission grid is viewed as a natural monopoly. This means that, if the goal is to promote competition among generators to sell electricity and among distributors to purchase it, then access to the transmission grid must be open on equal terms and conditions to all generators and distributors. (In Chapters 4 and 5, we consider the questions of how transmission prices might be set and whether generation, transmission, and distribution should be in separate companies.)

Independent system operator

At the beginning of this chapter, we referred to one of the peculiar features of electricity—that it cannot be stored, so total generation must always exactly equal total consumption—and noted that any proposal for implementing competition must accommodate the need to balance electrical loads. Under wholesale bilateral contracting, a new entity would be created, known as the independent system operator, or ISO. The ISO is responsible for balancing electricity demand and supply on the integrated transmission grid. It lines up emergency sources of electricity in case of unanticipated demand or equipment failure, and it has the authority to deny access to or shut down generators when unanticipated reductions in electricity use occur. This balancing has to be maintained on a virtually instantaneous basis to avoid blackouts or brownouts.

To ensure impartiality toward all of the generators that it is responsible for coordinating, the ISO is not allowed to own any generating facilities. The grid company could take on the ISO's role as long as the grid company does not also own any generating units. Because it is the sole provider of system balancing services, the ISO will be regulated, presumably by the Federal Energy Regulatory Commission.

To maintain the balance of supply and demand, as well as to maintain voltage levels on the transmission grid and to compensate for power losses that result from the physics of transmission, the ISO must have up-to-the-minute information on demands and supplies of electricity, as well as direct control over at least some portion of the generating facilities attached to the grid. Accordingly, the ISO enters into contracts with a select number of generating companies who make themselves available for immediate operation or shut-off. The ISO recovers the cost of these balancing services through fees charged to either the distribution companies or the generating companies.

Because electricity cannot be stored, electrical loads must be balanced so total generation exactly equals total consumption. Under bilateral contracting, the load-balancing function is performed by a new entity, the independent system operator, or ISO.

By charging penalty fees for load imbalances, an ISO minimizes its own need to resolve imbalances, as well as gives parties to a bilateral contract incentives to match their own supply and demand. Sally Hunt of National Economic Research Associates has proposed a variety of methods for setting these fees. One method, analogous to current practice in U.S. natural gas markets, consists of charging generating and distribution companies a substantial penalty for all load discrepancies; if the penalty is sufficiently high, it would encourage use of the spot market or some other means to avoid imbalances. Another method, currently practiced in the Norwegian electricity market, consists of charging the spot price for imbalances. Even though no centralized spot market is proposed under a bilateral contracting structure, informal spot markets are likely to arise as contracting parties look for ways to sell off their excess generation or to make up deficiencies in their own supplies, as well as to reduce their transaction costs.

The appropriate geographic scope of the ISO's service territory is yet to be determined. One option has a single ISO for each of the three highly integrated transmission grids that already exist in the United States. A more likely option, however, defines a territory coincident with each of the ten North American Electric Reliability Council reliability regions. (For a list of these grids and regions, see box in Chapter 2 on page 27.) Several of these regions, including New England and the Pennsylvania/New Jersey/Maryland region, already function as tightly integrated power pools with a centralized system operator. A third option, which may meet with favor among some state utility regulators and legislators, has the boundaries of ISO service territories correspond with state borders. This option is more likely if each state adopts its own approach to allowing more competition in electricity markets. Even with state-specific ISOs, the activities of the ISO are likely to be subject to regulation by the Federal Energy Regulatory Commission, which has jurisdiction over transmission pricing and all wholesale power sales.

Bilateral Contracting and Retail Competition

Under full retail competition, the ISO and the grid company play largely the same roles as in the wholesale version of bilateral contracting. The major differences between the wholesale and retail versions occur at the distribution and retail sales stages, where two new participants are added to replace the distribution company. The new participants and the variation in the transaction path are discussed next and illustrated in the figure on page 49.

Line companies

Under the retail competition version of bilateral contracting, the bundled power and distribution service provided by distribution companies is replaced with unbundled distribution service provided by regulated line companies. Line companies take high-voltage electricity from the transmission grid, lower its voltage, and distribute it to customers. These companies have a monopoly franchise to distribute electricity within their service territories and are subject to rate regulation by the state public utility commission.

Marketing companies

Marketing companies purchase electricity directly from generators and then sell it to customers, either bundled or unbundled with distribution services, which must be purchased from the line companies. Unlike the line companies, however, the marketers can be unregulated. With few barriers to entry into the marketing side of the business, this stage of the electricity industry is expected to be highly competitive.

Under retail bilateral contracting, the transaction path changes so that customers purchase electricity directly from generating companies or from competitive marketing companies at unregulated prices under terms specified in a bilateral contract between the two parties. Transmission and distribution services are obtained separately from the regulated grid and line companies, respectively. Both the transmission grid and the local distribution networks would be open to all generating companies and to all retail customers under equal and nondiscriminatory access rules. Retail access to generation markets would be phased in over time, most likely starting with the largest customers (industrial users) and working down to smaller customers (commercial

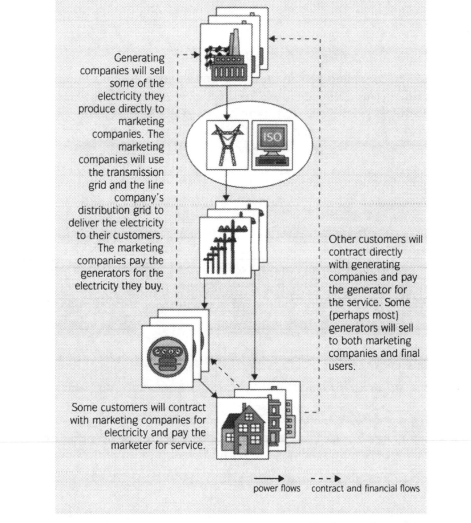

Retail
bilateral
contract
market
structure

Under retail bilateral contracting, distribution companies are replaced by *line companies*, which provide local distribution, and *marketing companies*, which purchase power from generators and sell it to customers. Marketing companies may or may not bundle their services with distribution.

Generating companies will sell some of the electricity they produce directly to marketing companies. The marketing companies will use the transmission grid and the line company's distribution grid to deliver the electricity to their customers. The marketing companies pay the generators for the electricity they buy.

Other customers will contract directly with generating companies and pay the generator for the service. Some (perhaps most) generators will sell to both marketing companies and final users.

Some customers will contract with marketing companies for electricity and pay the marketer for service.

→ power flows   ---▶ contract and financial flows

and residential users). During the transition to universal retail access, some portion of retail customers would continue to purchase electricity bundled with transmission and distribution services directly from their local line company at regulated rates.

The bilateral contract market structure appeals to many analysts, in part because it is so similar to current trading practices. However, if each of the different participants is indeed represented by a separate company, bilateral contracting represents a dramatic departure from existing practice. As we discuss in Chapter 5, some separation of functions may help ensure that generation markets are competitive. For example, it might be worthwhile to separate the transmission function from both generation and distribution to eliminate incentives for the grid company to provide preferential access to affiliated generating or distribution companies.

The Poolco Alternative: Wholesale and Retail

The major difference between bilateral contracting and Poolco market structures is that the Poolco model centralizes coordination of all short-run electricity transactions in a single official spot market known as the Poolco. (For more about spot markets, see box below.) Most power transactions would take place through the Poolco, except for some specific bilateral (known as *self-nominated*) contracts. Long-term contracts—referred to as *contracts for differences*—also would exist, but primarily to protect parties against risk associated with Poolco price fluctuations, rather than as specific bilateral contracts for a generator to supply power to a particular customer.

The participants and transactions of a typical wholesale Poolco model are illustrated in the figure on page 52; those for a retail Poolco model are shown in the figure on page 53. Under

What is a spot market? a contract market?

A *spot market* is a market where sellers have their commodities on hand and the goods are delivered immediately—on the spot. Spot markets are distinguished from *contract* (or *forward* or *futures) markets*, where the commodities are not physically present but, instead, are bought and sold via contracts that specify a delivery date in the future and fix a price.

Buying on the spot market is buying instantaneously; on the contract market, it is over the long term. The difference between the two is, for instance, the difference between buying a magazine at a newsstand (spot) or through a subscription (contract).

For some primary commodities, such as grain or oil, central spot markets have become established to facilitate large-scale trading; in electricity, many analysts envision that the Poolco would operate as something of a central spot market. It is also possible that a futures market in electricity could develop, as an alternative to contracts for differences and a means of hedging against price variability on the spot market.

expanded wholesale competition, the distribution companies buy from the Poolco; under retail competition, electricity customers and marketers are allowed to obtain power directly from the Poolco. Under a Poolco structure, the functions of most of the participants generally are the same as under bilateral contracting, but with some new and important participants and institutions, which we describe next.

Poolco

The Poolco is a regulated monopolist responsible for all of the functions assigned to the ISO under a bilateral contracting structure and for the operation of a centralized spot market for electricity. The Poolco solicits bids from generators regarding the prices and amounts of power they are willing to sell during each time segment of the following day (for instance, on a half-hour basis). Under wholesale competition, the Poolco also solicits purchase bids from distribution companies, specifying the projected amount of load or electricity demand at each segment during the day and the prices at which the distributors are willing to purchase the projected amounts of electricity. Under retail competition, a Poolco receives bids to purchase power from marketing companies and customers directly, instead of through distribution companies. Then the Poolco is responsible for finding the price per kilowatt-hour at which the supply amounts and bid amounts are equal. All generators receive the price in effect during the relevant period for the energy they supply. All power buyers pay the Poolco for the energy they purchase, plus a (regulated) surcharge to cover transmission costs and the Poolco's operating costs.

In most versions of the Poolco model, the task of coordinating the spot market and the task of operating the system are performed by a single entity, as we have assumed here. However, some proposals, such as the one adopted by the California Public Utilities Commission in December 1995, separate the two functions into two distinct entities: a system operator, often referred to as an ISO, who is responsible for operating the system and coordinating power delivery in real time, and a power exchange, which performs the market-making functions of the Poolco.

Self-nominated contracts

While most short-run electricity sales take place through the Poolco, transactions also can occur outside of the Poolco spot market. Generators and distributors (under wholesale competition) or generators and marketers or customers (under retail competition) may enter into long-term bilateral contracts for electric-

Wholesale Poolco market structure

Under a Poolco market structure, the coordination of all short-run electricity transactions is centralized in a single, official spot market, called the Poolco. Generators sell power in this spot market and distribution companies purchase it from this market. The Poolco also performs the functions of an ISO.

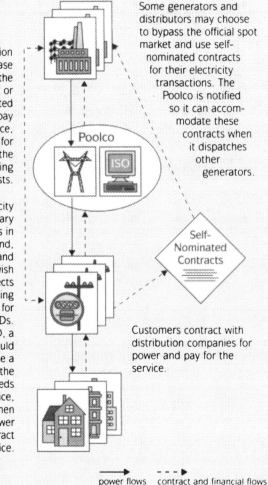

Distribution companies purchase electricity through the Poolco spot market or self-nominated contracts. They pay the spot market price, plus a surcharge for transmission and the Poolco's operating costs.

Because electricity spot prices will vary with fluctuations in electricity demand, generators and distributors may wish to mitigate the effects by establishing contracts for differences, or CFDs. Under a CFD, a generator could compensate a distributor when the spot price exceeds the contract price, and vice versa, when the spot price is lower than the contract price.

Some generators and distributors may choose to bypass the official spot market and use self-nominated contracts for their electricity transactions. The Poolco is notified so it can accommodate these contracts when it dispatches other generators.

Customers contract with distribution companies for power and pay for the service.

→ power flows   - - - ▶ contract and financial flows

ity supply. Under these long-term contracts, a generating company dedicates some portion of its production to a particular distributor, marketer, or customer. When such contracts, called self-nominated contracts, are written, participants are required to inform the Poolco so it can take the contract into account when it dispatches electricity from those generators who opted for centralized dispatch based on pool or spot-market prices. When contract supplies do not match actual demand by contract customers,

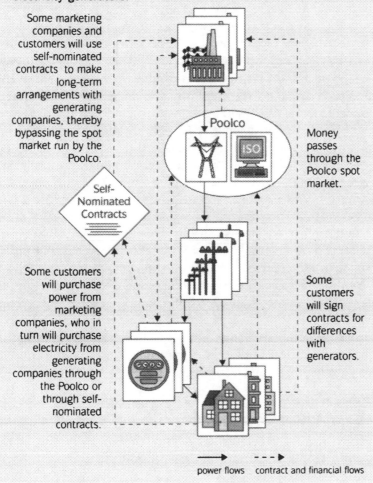

Retail Poolco market structure

Under a retail Poolco market structure, the distribution companies split into line companies and marketing companies, as with retail bilateral contracting. Generating companies will sell electricity into the spot market. They may sign contracts for differences either with marketers or with final customers. The Poolco will coordinate the spot market and dispatch of electricity generators.

Some marketing companies and customers will use self-nominated contracts to make long-term arrangements with generating companies, thereby bypassing the spot market run by the Poolco.

Poolco

Money passes through the Poolco spot market.

Self-Nominated Contracts

Some customers will purchase power from marketing companies, who in turn will purchase electricity from generating companies through the Poolco or through self-nominated contracts.

Some customers will sign contracts for differences with generators.

power flows   contract and financial flows

imbalances could be settled through sales to or purchases from the spot market.

Contracts for differences

Electricity spot prices will vary with fluctuations in electricity demand throughout the course of the day and across the seasons of the year. When demand is reaching the limits of available generating capacity, higher-marginal-cost generators are expected to come on line, and prices could be bid up quite high to mete out the available supply.

If generators or buyers are willing to pay to avoid exposure to price risks, then there are opportunities for profitable contracts between them. If generating companies are averse to risk, they might be willing to take a lower-than-average steady price rather than cope with the variation in spot prices over time. Similarly, a distributor, marketer, or customer averse to risk might be willing to pay a higher-than-average steady price to avoid the prospect of having to pay very high prices during peak demand periods. Consequently, there may be a range of prices—different from the average Poolco price—at which generators and buyers might want to contract with each other (or with financial intermediaries) to limit their exposure to the risk of sudden surges in the price of electricity.

Contracts for differences (CFDs) are the financial contracts that provide a means for generators and distributors who interact with the Poolco to avoid the price volatility that can result from spot-market transactions. To mitigate the risk of price variation, generators and distributors can sign contracts that specify a fixed price for electricity for some time period. Under these CFDs, sellers would compensate buyers when the spot price exceeds the contract price, and buyers would compensate sellers when the spot price is less than the contract price. Generators and distributors may be able to use futures markets and other independent trading institutions, along with or in place of CFDs, to hedge against the potential volatility of spot prices. However, CFDs also can include other payments between distributors and generators, such as compensation for any fixed costs of electricity generation that might not be recovered through sales in the Poolco spot market.

Comparing the Two Market Structures

The major difference between the bilateral contracting and the Poolco market structures lies in the scope of the activities performed by the system operators, referred to here as the ISO and the Poolco, respectively. Under bilateral contracting, given the power flows that have been arranged under bilateral contracts, the ISO balances electricity demand and supply by dispatching only a small number of the generating facilities attached to the grid. In doing so, the ISO pays the prices specified in contracts between the ISO and the generating companies. In Poolco settings, the Poolco dispatches most, if not all, of the generating units attached to the grid according to the quantity and price bids offered on the electricity spot market. Only those generating units that self-nominate out of the centralized pool—that is, that sign bilateral supply contracts with specific buyers—are not dispatched by the Poolco.

A variety of factors will affect the choice between the bilateral contracting or Poolco market structures or some hybrid of the two. These factors include the implications of each structure for effective load balancing, the time required to establish each type of structure, and the incentives for product differentiation under each market structure. We address each of these factors in turn below.

Is bilateral contracting consistent with effective load balancing and efficient use of the transmission system?

The main argument for favoring a Poolco over bilateral contracting arises from the need to adjust electricity supplies constantly to meet frequent and sometimes large fluctuations in demand during the course of the day. In the view of Poolco proponents, a single centralized entity provides the best way, if not the only way, to ensure that the power provided to meet customer demands is provided at the least cost. Under bilateral contracting, the ISO would provide similar system-balancing services and other transmission-related services (such as compensating for transmission line losses) through contracts with flexible generators. Assuming that its regulator provides the appropriate incentives, the ISO would dispatch those flexible generating units to balance the system at minimum cost. However, since the bulk of the electricity supplied in the market is traded under bilateral contracts, there could be times when a relatively high-cost generator owned by one generating company is operating while a lower-cost generator owned by another generating company sits idle.

If bilateral contracting results in large deviations from efficient dispatch of the electricity system and in inefficient use or pricing of the transmission grid, however, independent (rather than centralized) spot markets might well develop. These independent spot markets would offer generating companies a means to reduce the costs of meeting their contract obligations by letting them resell a low-cost generator's electricity during those periods rather than produce their own higher-cost electricity. Buyers also might resell electricity or accept payment for interruptions when other buyers are willing to pay extra for additional power. However, independent spot markets might be unable to offer the real-time precision necessary to achieve the lowest-cost outcome at any given moment. Unless the independent spot markets operate very rapidly, their use will go only part of the way toward eliminating system inefficiencies.

Would a Poolco delay the development of competition?

Those who prefer bilateral contracting to a Poolco suggest that establishing a centralized pool could delay, and perhaps prevent, the realization of the benefits of more competitive electricity markets. They point out that establishing a centralized pool takes much time and is costly compared to implementing a bilateral contracting system. Setting up a Poolco involves many potentially time-consuming activities, including defining its geographical scope, defining how it will be governed, and deciding how it will be regulated and by whom. On the other hand, the argument continues, a bilateral contracting system would build on current contracting practices in wholesale electricity markets. Therefore, once buyers and sellers of electricity obtained access to the transmission grid, they would be ready to execute trades among themselves using largely existing institutions. However, the process of setting up an ISO, including defining its regional jurisdiction, the rules governing its operation, and how it will be regulated, is also likely to be time-consuming and, therefore, the additional delay associated with establishing a Poolco may be short.

What are the incentives for product differentiation under each market structure?

The arguments in favor of Poolcos implicitly assume that all kilowatts of electricity are identical in the eyes of generators and buyers and that their only interest is the price at which electricity is exchanged. However, other aspects or qualities of electricity service, trading partners, or transactional arrangements could lead specific generators and buyers to sign bilateral supply contracts. For example, the customers in a particular utility's service territory could prefer that their electricity be generated using local coal; the local utility, therefore, might wish to sign a bilateral contract with a generator that burns local coal.

The scope for differentiation of electricity service is even greater under retail competition, where individual customers may have a range of preferences for attributes of generators beyond the preferences that may be held by distribution companies. For instance, customers could prefer electricity that has been generated with a renewable "green" technology, such as a wind or solar power. Certain industrial and commercial customers might be willing to purchase less-reliable electricity service if the price were right. Electricity customers also may want to purchase their electricity in the form of bundled energy services, that is, in combination with more-efficient heating, cooling, and lighting equipment.

Responding to such preferences is straightforward under bilateral contracting. For example, if a customer preferred renewable energy, it could express that preference directly by signing a bilateral contract with a generator using a renewable technology or with a marketer that has contracts with such generators. Opportunities for purchasing bundled energy services also are likely to abound, since marketers can sign fixed-price contracts with electricity generators and then bundle that electricity with the installation of energy-efficient cooling or lighting equipment at the customer's location. Arrangements for the purchase of interruptible power service also can be made directly using bilateral contracts.

All of these mechanisms are available under the Poolco, where generating companies and customers can bypass the Poolco through the use of self-nominated bilateral contracts. However, self-nominated contracts are not necessary for the provision of differentiated products. Generators and marketers also can use contracts for differences to offer differentiated services. For example, if a customer prefers power generated using local coal, it can sign a CFD with a generating company that is committed to burning locally mined coal. The contract would include provisions to assure the customer that the generator is actually generating whenever feasible and to assure the generator that it can earn enough revenue to cover its generating costs even if they exceed the market price. Marketing companies can use similar contract forms to affiliate themselves with particular types of generators. In another example, the consumer can sign a CFD with a generator that includes an agreement to be interrupted whenever the spot price reaches a certain level. To compensate the consumer for making itself available for interruptions, the generating company offers a low fixed price to the consumer. The generator can then sell the released power on the spot market when the spot price is high.

> *Some proposals assume that all kilowatts of electricity are identical in the eyes of customers, but are they? Factors other than price, such as the use of local coal or green technologies, may make some kilowatts more attractive than others.*

Poolcos and Market Power

Critics of the Poolco warn that a centralized pool may help to create *market power*—that is, the ability to raise price above competitive levels—in electricity spot markets, rather than facilitate true competition. The Poolco allegedly could exercise market power in two ways: it could create market power among the generating companies or, as a monopoly, it could exercise market power on its own behalf.

Could a Poolco create market power among generating companies?

One argument in favor of an independent Poolco that controls both the operation of the transmission grid and the dispatch of generators is that such an arrangement would remove the incentive for transmission companies to provide superior transmission access to affiliated generators without requiring that transmission companies divest themselves of generating capacity. However, if the Poolco structure limited the group of generators that could sell electricity into the pool, then it might prevent the pool from achieving the benefits to be derived from competition among generators. A highly concentrated generation market increases the chance that generators would collude to manipulate the spot price—that is, they could limit the amount of power sold in the market in order to raise the price and thus their profits.

To restrict the opportunities for such behavior, regulators need to be mindful of market shares and entry conditions when they are establishing Poolco territories and evaluating proposed utility mergers. When generation markets are concentrated (that is, have few participants), regulators should be willing either to increase the size of the pool or to permit entry from self-nominated generators outside the pool's region. If neither option is feasible (perhaps because of limits on transmission capacity or because of political constraints, such as state or international boundaries), allowing existing utilities to remain intact could result in concentrated generation markets. If the evidence on market concentration were sufficiently compelling, it could be necessary to consider breaking up existing generating companies into smaller companies.

Could a Poolco itself act as a monopolist?

As coordinator of short-term electricity sales, the Poolco facilitates the operation of the market but does not actually play a role as either a buyer or a seller in that market. In principle, the Poolco is a monopolist only in the provision of grid balancing and related services and in the operation of the electricity spot market. It is neither a monopolist (sole seller) in the supply of power to distribution companies nor a monopsonist (sole buyer) in the market for power supplied by generators. If, however, a Poolco is the sole gatekeeper to participation in an electricity market, it could exercise market power—charging high rates to power distributors and offering low prices to generators—as if it were a monopoly seller and monopsony buyer.

To ensure that the Poolco does not attempt to limit the number of transactions in the spot market, or charge excessively high rates for its services, or otherwise abuse its market position, regulation of the Poolco is necessary. A regulator must set up a method for pricing services supplied by the Poolco and be responsible for monitoring the Poolco's activities, primarily to ensure that the Poolco is truly facilitating all efficient transactions, given the bids of the market participants. Allowing market participants to bypass the pool through the self-nomination process also provides a check on the Poolco's abuse of market power. The California proposal to separate the system operator and spot-market operator functions into distinct entities also seeks to limit the ability of either party to exercise market power.

The debate over market structure for the electricity industry has yet to address adequately the questions of who will regulate the Poolco and by what method. Analysts who have offered opinions generally suggest that the Poolco should be allowed to collect revenues that cover the cost of operating the pool, including a reasonable rate of return on capital—in other words, traditional cost-of-service regulation. The Poolco's revenues could be collected as an add-on to the spot price of electricity. Rules of operation should maintain true independence on the part of the Poolco operator to ensure that the system is used efficiently and to preserve competition in generation markets. It may be necessary to require that the Poolco not be a part of any company that generates or markets electricity.

Conclusion

Legislative and regulatory reforms have already introduced some competition into the electricity industry. Technical change and current variations in price will only accelerate that trend. It is still an open question whether competition can be implemented satisfactorily just at the wholesale level or whether retail competition is necessary to achieve the cost and service benefits competition could bring. The appropriate structure for future electricity markets also is unsettled, with some observers arguing for broader use of bilateral contracts and others advocating a centralized spot market overseen by a Poolco. In either case, the competition must be implemented in such a way as to maintain the load balances necessary to preserve the integrity of the nation's electrical delivery system. Moreover, issues related to the pricing of transmission services must also be resolved; we turn to these in the next chapter.

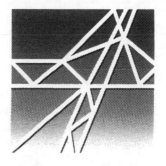

Chapter 4

Market Power in Delivering Power
Regulating Electricity Transmission

The collection of long-distance transmission lines crisscrossing the country—the *grid*—is the interstate highway system of the electricity industry. In the debate about the merits of deregulation and competition, nearly all participants presume that regulated access to power transmission lines and systems is necessary if the United States is to realize the potential for greater efficiency, reduced cost, and increased choices for electricity buyers. As we showed in Chapter 3, whatever approach to implementing competition and structuring the market is being considered, issues regarding access to and regulation of the transmission grid remain to be resolved. In this chapter, we consider those issues.

In April 1996, the Federal Energy Regulatory Commission (FERC) gave urgency to discussions about the transmission grid when it issued Order 888, describing the terms and conditions for open access to the transmission system. In this order, FERC requires transmission facility owners to allow nondiscriminatory access to all companies that wish to send electricity over their transmission lines for sale at the wholesale level to local distribution systems and other wholesale purchasers, such as power marketers. This open access must be accompanied by clear and specific tariffs for a variety of services and must allow for point-to-point and network-related charges, interruptible and noninterruptible (called firm in industry parlance) services, and a variety of ancillary dispatching and power management services. If the transmission facility owner also delivers its own electricity (that is, it is vertically integrated), it must implement *functional unbundling*—in other words, it must separate its different functions and charge itself the same rate for transmission that it charges unaffiliated generating companies. An important adjunct to the open-access policy is an *expansion obligation,* under which a transmission company must expand its capacity, if necessary, to customers willing to pay their share of expansion costs, unless the company cannot secure the permits necessary to construct additional facilities.

> *In April 1996, after more than a year of deliberation, FERC issued Order 888, which describes the terms and conditions for open access to the transmission grid for all companies wishing to sell electricity to local distributors and other wholesale purchasers.*

The major focus of FERC Order 888 and much of the debate about transmission pricing centers on ensuring nondiscriminatory access to transmission networks, rather than on pricing issues directly. More attention is being given to nondiscrimination than to pricing for many reasons. For one thing, to the extent that the companies using transmission lines that are subject to FERC Order 888 compete with one another, the prices they pay *relative* to their competitors have much more influence on their profits than do the *absolute* prices they all pay, which market forces (or regulators) tend to pass through to final users. For another, some industry observers expect that transmission costs would constitute only about 5–10 percent of the eventual price of electricity. (Other observers, however, foresee that transmission costs could be a much higher portion of electricity prices, particularly if transmission prices include fees for recovery of stranded costs; we discuss stranded costs in Chapter 6.)

Yet another reason for the emphasis on access over price is that state or federal pricing policies to control transmission company *market power* (that is, the transmission company's ability to raise the price above what it would otherwise be in a competitive market) will be futile unless nondiscriminatory access and expansion rights are ensured. Essentially, a transmission company could circumvent price regulation if it limited grid access to its own generators and then charged high prices for electricity. As a tactic, denying access ensures that independent generating companies could not enter the market and, through competition, bring rates down to levels closer to cost. In the other direction, preventing discrimination alone will do little good for end users if a transmission monopolist is free to charge everyone the identical monopoly price.

For these reasons, regulating price and regulating access go hand-in-hand. In this chapter, we look at the technical and economic factors that influence the decision whether to continue regulating transmission prices. We begin with three general principles about transmission pricing and then assess the need to regulate electricity transmission. Then, we consider conventional transmission regulation and present several alternatives. Finally, we examine four technical factors that greatly complicate electricity transmission pricing. In the next chapter, we explore in more detail the need for nondiscriminatory access and the options for achieving such access—specifically, the degree to which legislators and regulators might have to "restructure" electric utilities to bring about these results.

Premises for Regulating Transmission

Behind the presumption that regulation of transmission prices should continue are three underlying and related ideas.

- Continued transmission regulation will be necessary to promote efficient delivery of electricity from generators to distribution companies and, under retail competition, to marketers and end users. Because it is so costly for more than one company to have lines between a generator and a distribution company or customer, transmission is likely to remain a monopoly service. Competition among transmission companies cannot be counted on as a strategy to deliver electricity at reasonable terms and prices.

- Regulators can and will adopt methods for setting prices that lead transmission companies to respond to the present delivery needs of electricity generators and users, as well as to expand their facilities and adopt innovations to meet future demands for electricity delivery.

- The effort to regulate electricity transmission runs into complications beyond the usual demand and cost issues that face any regulator. Investments in transmission capacity and commitments to deliver electricity by one transmission company can affect how much it costs *other* transmission companies to deliver electricity. Such side effects may have implications for decisions about the ownership of transmission facilities, the feasibility of competition, and the determination of which facilities should be regulated.

To keep these complexities manageable, we can regard these three issues as posing three questions: *whether* to regulate transmission, *how* to regulate transmission, and *which* transmission facilities to regulate.

Answers to these questions are crucial. Transmission prices that are too high could lead to the wasteful use of forms of energy other than electricity, to excessive cogeneration, and to placement of power plants closer to end users than is necessary. High transmission prices that make it uneconomical to ship electricity long distances may preclude the kind of competition among generators that restructuring is intended to promote. If regulators set transmission prices too low, however, utilities will have too little incentive to install, maintain, and expand the capacity necessary to ensure that the rest of the electricity industry can perform competitively and efficiently. Moreover, new generating companies then may choose to locate farther away from users than is genuinely economical.

Whether To Regulate Transmission

Proposals to extend wholesale or retail competition into electricity markets are driven by the desire to introduce market information and incentives to an industry that has been largely directed by regulators. Many observers agree that electricity generation can be left to the market—perhaps with restructuring to ensure effective separation of generation from transmission, as we discuss in the next chapter. There is far less optimism regarding competition in the transmission of electricity between generators and end users, local distribution companies, or other adjacent transmission companies for further delivery. The prevailing belief among industry observers seems to be that regulators will determine the prices and conditions for transmitting electric power for some time to come. This premise should not be adopted, perhaps, without deeper examination.

What is a monopoly?

A *monopoly* is a market with only one seller. By being the sole supplier of an item and facing no competition, a monopolist can select the price that is most profitable for itself. For many people, the harm of monopoly is primarily the transfer of wealth to a monopolist from the buyers who have to pay the monopolist's prices. Economists emphasize the inefficiency that follows when a monopolist raises its price, because to do so it generally must reduce overall output in its market below what would be sold at lower, competitive prices.

Monopolies can be public when the government is the sole supplier; the U.S. Post Office and municipal transit, water, and electricity distribution companies are examples. In the United States, though, most monopolies are privately owned. Whether private or public, monopolies are sometimes created by the government through laws and rules that limit entry and competition. Firms may attempt to create monopolies through collusion, but antitrust laws exist to prevent such conduct. When technology and capital costs imply that one firm can provide a good or service at less expense than two or more firms—that is, when a *natural monopoly* exists—those markets may come to be monopolized as well. In all of these cases, the government may act, sometimes for good reason, to regulate the prices these monopolies can charge.

The presence of natural monopoly conditions, however, does not necessarily imply the absence of competition. If the cost advantages for a single firm are not great or if competition is not especially intense, multiple competitors may be able to survive in a market. Alternatively, a firm with a natural monopoly may feel pressure from potential entrants that want to replace it as the monopolist in that market. Finally, the monopolist may face competition from suppliers of other commodities that can be substituted for it. For example, an electric utility may face competition from other fuels (such as oil or natural gas) for some uses (such as home heating).

Electricity transmission is generally presumed to be a *natural monopoly* service—that is, one in which a single supplier can serve the entire market at lower cost than two or more firms can—and therefore to require regulation. (For more about monopolies, see box above.) The presumption of natural monopoly alone, however, is not enough to justify the administrative costs and inefficiencies that inevitably accompany regulation. In general, for regulation to be politically and economically worthwhile, consumers must face losses from monopoly pricing that are large enough to cancel out the costs of regulation. This condition holds true for products that most (if not all) people purchase in substantial amounts and that people would keep purchasing in nearly as large amounts, even if the monopolist charged prices that were significantly, if not exorbitantly, higher than the monopolist's costs to provide the product. Under such considerations, competition is least feasible—and regulation is most likely to be worthwhile—in large markets where fixed costs are a high proportion of the total cost of serving the market and cannot be recovered once they have been sunk.

Electricity transmission does seem to exhibit the characteristics of a monopoly requiring regulation. In the first place, electricity is clearly a sector important to the economy and a product that, for many uses, has no economical substitute. Second, the physical aspects of electricity transmission are such that a single supplier may be able to serve an entire market at lower cost than two or more firms could. Substantial fixed costs result from the construction of the wire networks for local distribution grids or long-distance transmission lines. Once these networks are installed, it is typically less expensive to expand capacity by adding lines or amplifiers along those lines than it would be to start new networks from scratch.

At least in theory, transmission companies may face potential competition in indirect ways. For instance, suppose an electricity consumer is connected to more than one independent transmission network. It then could purchase electricity from generators connected to different transmission networks. Alternatively, suppose that a generating company can ship electricity through different transmission networks to different groups of users. In either case, if one transmission company raises prices or denies access to these customers or generators, they can turn to other generators to obtain or deliver electricity.

The interconnected nature of transmission networks, as we discuss later in the chapter, may limit the possibility of having multiple independent connections available to any particular generator or customer. However, over the longer term, other possibilities may indirectly limit the market power of transmission companies. New producers of electricity could construct their generators close to towns that are subject to high-priced transmission service, so that electricity is transmitted across shorter distances. Large industrial users could produce some of their own electricity rather than using the transmission lines. Industrial users also could be willing and able to relocate their energy-intensive operations to regions where electricity rates, including transmission charges, are lower. Technological change, such as improvements in solar power, may make it economical someday for individual residential and commercial customers to generate their own electricity on site, eliminating the need for long-distance power transmission. It is important to note, however, that relocation and self-generation of power as alternatives to an existing transmission grid could be inefficient to the extent that users turn to them only because transmission rates are at monopoly levels.

How To Regulate Transmission

Assuming potential competition does not provide a sufficient check on the market power created by natural monopoly in electricity transmission, we might expect regulation to continue. The basic goal of regulation is deceptively simple: set prices as low as possible, consistent with raising enough revenue to cover the utility's costs, including a fair return on its investment. As we noted in Chapter 2, the traditional way to achieve this has been what is called cost-of-service or rate-of-return regulation. Because the traditional approach to regulation has some limitations, several alternatives have been developed. In this section, we discuss both traditional regulation and some of the most promising alternatives.

Conventional Rate-of-Return Regulation

In its simplest form, rate-of-return regulation of an electricity transmission company is computed in three steps:

1. Estimate the amount of electricity (kilowatt-hours, or kWh) the transmission system will be delivering over a given year.
2. Estimate how much it will cost the transmission company to transmit that quantity of kWh.
3. Divide the second number by the first to compute a transmission price in cents/kWh.

Behind this simple calculation lie some serious problems associated with information, implementation, and incentives. Rate-of-return regulation requires regulators to ascertain demand and costs and to set prices so that investors can earn an appropriate rate of return on the money they have tied up in regulated companies. This is not an easy task.

Demand for transmission may be relatively easy to forecast from current usage patterns if it is relatively insensitive to price—that is, if demand for electricity doesn't go up a lot when the price goes down, and vice versa. Changes in technology, population, income, or weather that affect electricity use, however, make demand forecasts less reliable. This unreliability introduces a *downside* risk that there will be insufficient demand to cover transmission costs and an *upside* risk that capacity will be insufficient to meet demand, with resulting reductions (brownouts) or interruptions (blackouts) in power delivery.

More serious complications exist on the cost side of the regulatory equation. If the utility knows that the higher the regulator believes its costs to be, the higher the price it can charge, it has an incentive to exaggerate costs, pocketing the difference. But, paradoxically, even if the regulator *can* determine accurately costs and set prices accordingly, the regulated utility still does not profit from minimizing operating costs, using equipment efficiently, or supporting research leading to innovations that would reduce costs or benefit consumers in the future. Therefore, effective rate-of-return regulation requires that regulators exert considerable micromanagement of the operations of the utilities they regulate. It commits the government to a continuing need to investigate and audit the cost data reported by utilities, creating the potential for disallowances and protracted legal disputes.

Alternatives: Incentive Regulation and Joint Ventures

As we noted in Chapter 2, problems with rate-of-return regulation have led to the development of alternative, incentive-based regulatory schemes that attempt to divorce price from cost. It could be argued that traditional rate-of-return regulation has some informal incentives already built in, in the form of *regulatory lag*, the gap of time between when a regulated utility lowers its costs and when the regulator orders a corresponding rate reduction. This gap gives the regulated utility an incentive to reduce its costs and "profit" by doing so for some period of time.

An alternative to rate-of-return regulation, the price-cap approach is in many respects an extended and formalized version of regulatory lag. It came to prominence in the telecommunications industry as a way to regulate the long-distance services of AT&T following its divestiture of its regulated monopoly local telephone companies. In this case, incentive regulation was attractive partly because growing competition in long-distance service was expected to keep AT&T's prices from diverging very far from the regulatory ideal.

In theory, price caps work by having the regulator set an initial price that is at a level reasonably close to current costs. This initial price is then allowed over time to adjust for inflation. Once the initial price has been set, future prices are independent of the utility's actual costs, thus creating an incentive for the utility to economize, innovate, and become more productive. To share the expected benefits of this increased productivity with customers, the utility and its regulator agree in advance on a productivity factor, usually a percentage by which prices after inflation would be reduced annually. The resulting price reduction gives users some of the benefits of the increase in productivity brought about by incentive regulation. It is important, however, that this productivity adjustment factor be fixed in advance. If it is allowed to reflect the utility's actual cost reductions and innovations, it strips away the incentive to become more productive.

For price caps—and other incentive-based regulation schemes—to succeed in practice, it is crucial that all sides commit to prices *in advance* and stick to them, regardless of how productive and profitable the utility actually becomes. If the utility—in this case, a transmission system—turns out to be more profitable than predicted, because of unanticipated success in reducing costs or in coming up with bold innovations, there will be political pressure to reduce the utility's rates, transferring some of those profits to consumers. In the other direction, a utility that ends up losing money because its productivity did not increase as fast as expected undoubtedly will seek rate relief from the regulators. The utility's case is likely to have considerable force, because of legal doctrines requiring regulators to allow utilities the opportunity to earn "just and reasonable" returns. These doctrines are not merely a matter of fairness; they help promote the funding necessary to provide regulated services by ensuring that the government will not refuse to compensate investors after their investments have been sunk.

Problems with traditional rate-of-return regulation have inspired new ways of regulating retail electricity rates that provide incentives for utilities to behave efficiently. Incentive-based methods include price-cap, yardstick, and sliding-scale approaches.

Regulators and industry experts have proposed other methods to provide incentives to regulated firms to control costs. A sliding-scale or profit-sharing approach, in which firms keep profits within a band around a target rate of return but have to reduce their rates if profits exceed the band, balances cost-cutting incentives with the political constraints that set limits on a regulated firm's profits. Yardstick approaches, which base rates on the prices or costs of similar firms in other markets, allow a regulator to tie rates to some information about costs, while allowing the firm it regulates to profit from reductions in its own costs relative to the costs of similar firms.

A final alternative to traditional rate-of-return regulation is the *competitive joint venture* approach. This method would allow a number of independent firms long-term or permanent partial ownership rights to a share of a transmission line or system. Assuming such a transformation of the current ownership structure were possible, competition among these part-owners would lead them to provide access to their transmission to buyers at lev-

The long-distance transmission lines that crisscross the United States are the interstate highways of the electricity industry.

els close to cost. One difficulty, explained in more detail below, is that a contract to deliver power over one's transmission lines does not guarantee that the electricity will actually flow over those lines. Moreover, for these joint ventures to lead to competitive performance, each owner must have the right to expand the transmission grid. In circumstances when the costs of this expansion are less than the cost of initial acquisition of rights-of-way and construction, potential sellers of transmission service may want to wait and expand an existing line rather than contribute to its initial construction. Consequently, regulation may still be required to ensure that those who wish to expand capacity make cost-based contributions to the initial owners toward the initial construction costs. However, even if regulation of capacity rights is required, the ability to forbear from regulating day-to-day or minute-to-minute transmission rates could make transmission markets more efficient.

Making Pricing More Efficient

So far in our discussion, we have regarded electricity transmission as a single service that is sold on identical terms and conditions to all customers, but this is not necessarily the way things work in the marketplace today. Electricity is sold under a variety of terms and conditions to a variety of users, a practice that creates the potential need for different prices to reflect different circumstances and costs.

Perhaps the most widely recognized factor for varying price in the electricity industry is that between high-demand, or *peak*, periods and low-demand, or *off-peak*, periods. While peak-load pricing is often applied to generation, it may apply to transmission as well. During peak periods, when capacity is used most intensively, a transmission company's cost of delivering an additional kilowatt of power will be the cost of the extra lines and transformers it has to build in order to do so. During off-peak periods, the cost of delivering an additional kilowatt of power will be lower, since the transmission company can use the capacity it already has in place. In practice, electricity use may be subdivided into several demand-differentiated periods, depending on the time of day, the season of the year, economic business cycles, and random fluctuations in the weather, each with potentially different rates appropriate for efficient use of the transmission network.

Another factor relevant for costs and pricing is distance. The costs of rights-of-way, lines, and power losses are roughly proportional to distance. The longer the distance a transmission line must cross to deliver electricity from a particular generator to a particular end user or local distribution grid, the more costly will be the transmission, all else being equal. However, basing transmission pricing on distance, rather than using a postage-stamp model where pricing is independent of distance, may spark objections from customers and users located far away from generators.

Differences in demand for electricity are yet another factor that may influence pricing practices. For instance, one consideration in setting transmission prices is whether the service is interruptible or firm. Interruptible service may be priced lower according to the willingness of a generator and its customers to accept interruptions in power delivery and the notice they require prior to those interruptions. Other customers, however, must have firm commitments that the power they need will be delivered when they need it. A second, related consideration is the degree to which the buyer's demand for electricity varies over time. Variability of power demand, unwillingness to accept interruptions, or unwillingness to make long-term commitments for power delivery can increase the costs of ensuring that sufficient transmission capacity exists to meet the customer's electricity needs at any given time. Finally, principles of efficient regulatory pricing suggest that, to minimize the costs associated with stifling demand, regulators should set rates relatively higher to those customers less willing to cut electricity use in response, and keep rates closer to cost for customers whose demands are more sensitive to changes in price.

If regulators accept that electricity transmission is sold to a variety of users under a variety of terms and conditions, then various methods for improving the economic performance of regulation become available. One such method is the *two-part tariff*, a charge for transmitting electricity that comes in two steps. The customer pays, first, a flat fee for the right to transmit electricity and then a per-unit price for each kilowatt-hour of energy transmitted. Two-part tariffs are akin to volume discounts, where the user pays a high price for the first few kilowatts transmitted and a lower price for subsequent power deliveries. The advantage of two-part tariffs is that, when some of the cost of the capital investment for the transmission lines is covered via fixed fees, then regulators can reduce the per-unit price of transmission to a value closer to the cost of providing the last unit of power they

What is marginal cost pricing?

The *marginal cost* of a good or service is how much it costs to produce just one more unit of it. The marginal cost is derived from the variable costs—the extra labor and raw materials, for example—and does not include fixed costs, such as the capital cost of the factory. The marginal cost of the just-one-more item may go up, go down, or stay the same, compared with the previous item.

Marginal cost pricing refers to the occurrence, in a competitive market, of selling prices being equal to the marginal cost. Most people think this means that a company explicitly identifies its marginal cost and sets its selling price accordingly. An economist, however, sees the process unfolding differently: a company identifies the price that the market is willing to pay and then produces an item as long as its marginal cost is *less* than the selling price. Once the marginal cost surpasses the selling price, the company loses money by making just one more. Therefore, the company stops producing at the point where its marginal cost equals the selling price.

demand—what economists call *marginal cost*. (For more about marginal cost pricing, see box on page 72.)

Two-part tariffs are not without their disadvantages. The fixed fees may drive some customers away, especially if they would not purchase enough transmission to make paying the fixed fee worthwhile. Fixed fees also invite potentially inefficient consolidations of buyers who band together solely to avoid paying the fixed fee more than once. These potential disadvantages are less serious if the fixed part of the two-part tariff reflects the up-front cost of building transmission capacity to serve a particular customer, with the per-unit charge covering the cost of using that capacity to deliver electricity.

Special Features of the Transmission Industry

The preceding points apply to price setting in any regulated industry and, in particular, to local distribution of electricity, which also is likely to face continued regulation. Policymakers, however, will have to take into account special technological circumstances that further complicate transmission pricing and regulation. We now turn to four of the most crucial issues raised by the special circumstances of transmission technology: balances between generation and use, power losses and congestion, a phenomenon called "loop flow" and the associated network effects across interconnected transmission grids, and specific asset issues.

Balancing Generation and Use

As we have already noted, electricity cannot be stored or compressed in the way that other delivered energy sources, such as oil and gas, can. Consequently, what goes into a transmission network can be no more than what the customers want to take out. If brownouts and blackouts are to be avoided, the grid must carry as much electricity as the customers demand at the prices they have to pay. Inability to store electricity and the need to ensure its availability together imply that generation and use have to be equal all of the time.

While it is still too early to know who among the participants—transmission companies, local independent system operators, Poolcos, or the buyers and sellers themselves via contracts—would best address load imbalances, we may speculate on what would happen if the responsibility for load balancing falls at the transmission link of the electricity chain. In this case, transmission access contracts could require that, when a generator's customers demand less electricity than the generator is producing, then the generator must either cut back its own output or have a contract with another generator who agrees to cut back its output. If customers demand more electricity than the generator is producing, transmission contracts could specify that the generator is obliged to provide additional electricity or to grant the operator of the transmission network the right to reduce or interrupt service to that particular generator's users.

In setting prices for the transmission industry, policymakers must take into account special technological circumstances of electricity transmission: load balancing, line losses, loop flow, and user-specific lines.

To address the load-balancing issue, transmission rates charged to generators who ship electricity through their lines should reflect the expected imbalance costs the generators and their customers will place on the transmission grid as a whole. To the extent that generators, local distribution companies, and customers either commit to a firm level of power generation and consumption or take on the liability for divergence from the optimum, transmission prices to them could be lower. Those firms that shift the liability onto the transmission company will pay higher prices for transmission. As a last resort, the operator of the transmission network should retain the ability to cut off customers or refuse to take power from certain generators whenever unforeseen contingencies or breaches of load-balancing contracts result in imbalances that no one else will correct.

Power Line Losses

No physical transmission line is a perfect conductor of electricity. Because all substances, to some degree, have *resistance* (that is, they resist the flow of electrical current), some electrical energy sent through a line is always lost as heat. In terms of transmission costs, resistance means that a customer who purchases the right to transmit X kilowatts through the system will take out only $X - Y$ of them, where Y is the number of kilowatts lost due to resistance. In effect, this creates a cost of transmission that the customers—generators and wholesale or retail electricity purchasers—bear. All else being equal, these costs are likely to be proportional to the distance traveled, although loop flow effects described below may mitigate the costs to some extent. According to the Electric Power Research Institute, a typical transmission and distribution system loses 10 percent of its energy through losses to resistance.

At sufficiently high power levels, a transmission line begins to become congested. The additional electricity the line can deliver becomes less than the additional electricity that generators supply. The grid reaches the point where attempts to send additional power through the line cause thermal breakdowns, akin to blowing a fuse at home. This determines an effective capacity limit for the transmission line.

In the traditional electric utility, load loss and line capacity are managed through systems engineering, and service is curtailed during emergencies. With separate, unbundled transmission, regulators will have to decide whether the transmission charge should be based on the quantity of energy that power generators put into the system or on the amount that the distribution companies and end users take out. In addition, load losses may depend not only on how much energy the line is carrying but on the rate at which the energy is being put into the line. Efficient transmission pricing will have to take both into account.

Loop Flow Effects

While an easy and useful image of a transmission line is a single cable connecting a generator to a local distribution system or end user, in practice the situation is much more complicated. A grid of interconnected transmission networks covering large regions within the United States constitutes a more accurate picture. Utilities interconnect their systems to facilitate energy exchanges, manage load imbalances, and improve system reliability. Transmission systems also interconnect when businesses and population expand into new areas. When access to the transmission grid is made more open and electricity generation markets then expand to cover wider areas, the crisscrossing of interconnected transmission lines will look even more complicated than it does today.

As noted above, a theoretical benefit of the interconnection of separately owned transmission networks is that it could facilitate competition in transmission. However, interconnection brings with it a side effect, peculiar to electricity transmission, that with current technology makes it virtually impossible to operate interconnected networks independently. This phenomenon, known as loop flow, thus has a direct bearing on the necessity of regulation and how transmission prices can be set.

It is the nature of electricity that, to get from one point to another, it makes use of all available routes to get there. These flows are called "parallel" or "loop" flows, and they occur because every time a new parallel route is opened between two points, the electrical resistance between those two points is reduced. Pumping water offers a useful analogy, because adding a second pipe between two points increases the capacity of the system and so reduces the amount of force necessary to move the water to its destination. The water analogy is imperfect, however, because electricity flowing across transmission lines has no equivalent to a shut-off valve that regulates the flow of water. The figure on page 77 gives a fuller explanation of why loop flow happens, and we refer to the diagram throughout the following discussion.

Loop flow has a most definite effect on the pricing of transmission services. If Transmitter 1 agrees to transmit more electricity from Generator 1 to Customer 1, some of that electricity will flow over Transmitter 1's lines; however, because the transmission lines are interconnected, some of the electricity *also* will flow over Transmitter 2's lines (between points C and D). The electricity flowing over Transmitter 2's lines will increase Transmitter 2's cost and load loss on its lines between Generator 2 and Customer 2.

It is even possible that, depending upon line capacities and resistance, Transmitter 1 could agree to transmit more power between Generator 1 and Customer 1, and yet most of that power actually would end up flowing over Transmitter 2's lines. In the jargon of the industry, the *contract path*—that is, the route nominally specified in an agreement to have electricity transmitted between two points—may differ from the *power path*—the route electricity takes over the transmission grid. In practical terms, Transmitter 1 could pocket the transmission fees for delivering the additional electricity between Generator 1 and Customer 1, while Transmitter 2 ends up bearing some of the actual costs of that transmission.

Because of loop flow, a transmission company does not capture the full benefits of expanding its own capacity nor does it bear the full costs of increasing its contractual obligations to transmit power. In our example, if Transmitter 2 increases its capacity to transmit electricity between Generator 2 and Customer 2 by adding lines between points C and D, the amount it costs *Transmitter 1* to deliver electricity between Generator 1 and Customer 1 will fall. More of the electricity that Transmitter 1 agreed to transmit between Generator 1 and Customer 1 will go over Transmitter 2's expanded line, reducing Transmitter 1's load losses and other costs associated with using its own facilities. In short, Transmitter 1 benefits from Transmitter 2's capacity expansion.

Loop flow effects complicate the capacity, congestion, and pricing issues associated with electricity transmission. For example, suppose it is less expensive to produce power at Generator 2 than at Generator 1. Because of loop flow, increasing the capacity of Transmitter 2's lines by one megawatt may allow generators at Generator 2 to produce much more than one additional megawatt

The figure shows a simple diagram of an electricity network, with two generation companies (Generator 1 and Generator 2), two transmission companies (Transmitter 1 and Transmitter 2) and two customers (Customer 1 and Customer 2), who are, perhaps, local distribution grids for a city. The points A, B, C, and D indicate junctions where the grids for Transmitters 1 and 2 are interconnected.

For the purposes of our illustration, suppose that Transmitter 1 owns the transmission line that runs directly between Generator 1 and Customer 1; suppose that Transmitter 1 also owns the interconnecting line that runs vertically between points A and C. (In the figure, Transmitter 1's lines are shown in blue.) Suppose that the other transmission company, Transmitter 2, owns the transmission line that runs directly between Generator 2 and Customer 2, as well as the interconnecting line between points B and D. (Transmitter 2's lines are shown in black.) Transmitter 1 has a contract to deliver electricity from Generator 1 to Customer 1, and Transmitter 2 has a contract to deliver electricity from Generator 2 to Customer 2.

Loop flow is the name given to the characteristic of electricity that it takes all available routes to get from one point to another. In a very simple example, this means that, say, if a second line becomes available that is identical to the first, then the electricity that had been flowing over the first line will "divide" itself, so half of it will flow over the first line and half over the second. The electricity being transmitted doesn't "know" that it is supposed to stay on the lines owned by the company offering the transmission service.

The implications of loop flow effects for pricing transmission

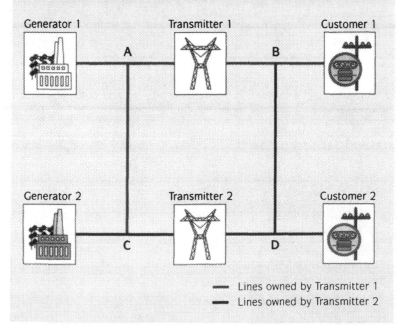

Generator 1   Transmitter 1   Customer 1

A   B

Generator 2   Transmitter 2   Customer 2

C   D

— Lines owned by Transmitter 1
— Lines owned by Transmitter 2

of power, displacing that much power from the more expensive Generator 1. Exactly how much more power this allows generators at Generator 2 to supply depends on the physical properties of the interconnected transmission grids. The economic value of that displacement depends on the costs of generating electricity, which can vary considerably over time because of variations in demand and supply throughout the system. Calculating the economic benefits from increasing Transmitter 2's capacity should take these intricate displacement and cost effects into account.

Designing mechanisms to tell regulators and transmission companies when it is economically worthwhile to expand capacity thus becomes a highly complex task. Among recent proposals for dealing with loop flow effects associated with capacity and congestion are methods for computing "nodal" prices at each input or output point on a transmission grid. Proponents of this method argue that it will encourage generators and their customers to recognize the effects of their demands on overall network capacity and to signal where the transmission grid's capacity is most in need of expansion. To encourage transmission companies to reduce load losses and congestion by increasing grid capacity, regulators should recognize all of these factors in deciding whether and how to implement incentive regulation, two-part tariffs, or customer-class specific prices.

Industry experts have noted that the pricing problems caused by loop flow will become greater as increased competition causes more generators to want to deliver power to more destinations. In the effort to address loop flow, the transmission industry could informally or formally consolidate into large regional companies, perhaps with only a few covering the nation. Alternatively, a single independent system operator could manage several interconnected grids owned by separate transmission companies. This consolidation, however, could delay or deny any potential benefits of competition among transmission companies. In the final analysis, the loop flow effect could become the primary reason why the need to regulate transmission will persist.

Specific Assets and the Definition of "Transmission"

If loop flow effects create a need for coordinated transmission pricing and capacity decisions and also increase the likelihood of continued regulation, spreading the benefits of competition will require that we look for ways to limit the scope of this transmission monopoly. A similar issue was faced in the telecommunications industry in conjunction with the breakup of AT&T. Where did the local telephone monopoly end and the competitive long-distance market begin? Even though the application in the telecommunications industry has been and continues to be quite controversial, the underlying principle is sound: assets devoted to providing service to a particular user should go with (and be paid for by) that user; only those assets that users cannot economically provide for themselves should be regarded as part of the regulated monopoly.

The loop flow figure we have used so far can be used again to illustrate some of the considerations that could play a role in allocating assets between the transmission sector and the other sectors of the electricity industry. Consider the line connecting Generator 1 with the transmission grid at point A. If Generator 1 is the only generator along that line, we might regard that line as specific to Generator 1, and thus part of Generator 1's facility. A similar argument might rationalize allocating lines such as that between point D and Customer 2 to Customer 2's distribution system. If this approach is carried to the extreme, then perhaps the only assets that would have to be regulated as transmission are those that are part of the system bounded by points A, B, C, and D. Weighing against this allocation would be potential economies that come from having the transmission system install and operate the user-specific lines; for example, there could be savings from coordinated maintenance of user-specific lines along with the main grid. Moreover, if more than one buyer or seller connects to the user-specific lines (for example, if there are many generators at the point where Generator 2 is located), then the line may not be specific to a particular competitor and, thus, may be better treated as part of the transmission grid for regulatory purposes.

Conclusion

All the factors that affect the pricing of electricity transmission need to be considered in connection with the different jurisdictional responsibilities held by FERC and the state public utility commissions. Ideally, loop flow effects would be contained within the transmission facilities subject to FERC jurisdiction, and the assets under state jurisdiction would be those specific to particular generators or delivery points, such as local distribution systems. In fact, FERC concluded in Order 888 (which we mentioned earlier in this chapter) that any facilities used to deliver electricity to wholesale resellers are subject to its jurisdiction, including price regulation and open-access rules. This leaves states to regulate facilities that deliver power from wholesale resellers to their final customers. When a utility's facilities are used not for resale but to deliver power under contract from a generator to an end user, FERC claims that it has jurisdiction over at least some of the facilities reaching all the way to that end user, however.

The analysis of legal jurisdiction is likely to turn on interpretations of relevant statutes and the meaning of "interstate commerce," as well as on political debates among FERC, state regulators, and the utility industry. To some extent, the economic considerations associated with load balancing, line losses, loop flows, congestion pricing, and specific assets may influence the outcomes of these debates. Perhaps more importantly, the optimal design of policies to deal with these considerations will undoubtedly need to take into account the legal authority and political ability of policymakers to manage transmission networks as competition in the electricity industry grows.

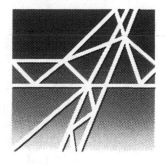

Chapter 5

Restructuring Electric Utilities
The Pros and Cons of Vertical Integration

In Chapter 3, we looked at the ways in which competition will be promoted and implemented in the electricity industry, beginning among generators seeking to sell power at the wholesale level to utilities and perhaps extending at the retail level to industrial users and, through marketing companies, to commercial and residential users. Then, in Chapter 4 we examined the transmission of electricity, which appears likely to remain a regulated monopoly for the foreseeable future. We focused broadly on pricing but we noted that, from a regulatory standpoint, pricing issues are most prominently presented in the context of nondiscriminatory, open access to transmission networks.

In this chapter, we show that the attention to discrimination is warranted

81

because providers of transmission services who are regulated and own generating capacity may have incentives to favor their own generators with better access—that is, with the ability to transmit efficiently at the regulated rates, while everyone else may face delays, interruptions, or greater power losses in transmission services. In the current, regulated scheme of things, many electric utilities perform all four functions—generation, transmission, local distribution, and retail sales—under some sort of government oversight. But if competition is expanded for the generation segment of the industry while transmission and distribution continue to be regulated, then electric utilities would find themselves in the position of having some parts of their companies regulated and others competitive.

As we showed in Chapter 2, back in 1935 the government intervened through the Public Utility Holding Company Act in part to correct abuses that were perceived to be the result of electric companies' having one foot in the competitive sector and the other in the regulated sector of the economy. Since the initiatives now being contemplated to promote competition may reproduce the same situation for electric utilities today, we review the advantages and disadvantages when utilities operate at multiple levels within the electricity industry—what economists refer to as *vertical integration*. Can efficient, nondiscriminatory access to transmission lines be achieved by open-access rules alone? Or, instead, will more severe measures be necessary, such as a requirement that generation and transmission be provided in fully separated subsidiaries? In the most extreme view of the case, policymakers may need to consider a divestiture that would place the regulated monopoly services and the potentially competitive services in completely different corporations, as was done with AT&T in 1984.

If competition is expanded for the generation segment of the electricity industry while transmission and distribution remain regulated, utilities could wind up with some parts of their business regulated and others competitive.

The fact that "restructuring" has come to be synonymous with expanding competition in the electricity industry underscores the prominence of these considerations. The form and extent of restructuring will depend on whether the potential competitive benefits of imposing functional or structural separations between the regulated and unregulated sectors of the electricity industry—including, for example, the creation of independent system operators or of Poolcos as described in Chapter 3—outweigh the benefits of continuing the traditional, vertically integrated structure of electric utilities.

To explain the vertical restructuring issues facing the electricity industry and policymakers, we first offer some general observations about how industries are structured and review four justifications that are frequently cited for vertical integration. Then, we look at vertical integration as it operates in regulated markets, noting particularly potential abuses when integration straddles the line between regulated and competitive industry sectors. Finally, we connect what we know about vertical integration with key issues in the debate about restructuring the electricity industry.

Pros and Cons of Vertical Integration

The production of virtually any good or service ultimately destined for consumers involves many steps in a chain, beginning with labor and natural resources at one end, proceeding through different manufacturing and development stages, then through wholesaling, distribution, and eventual retail sale to the customer. Steps along the chain of production usually are described as being *vertically* related, as distinct from *horizontal* relationships among firms at the same stage of production. In the electricity industry, for example, a generator and the coal mines that supply it are vertically related, while the coal mines themselves are horizontally related to each other. A useful way to understand horizontal and vertical relationships is this: horizontally related firms compete against each other (like the coal mines), while vertically related firms operate in buyer/seller relationships (like the generator and each coal mine). "Vertical" also is used to describe relationships between firms that provide products that are used together (bread and butter) rather than substitute for one another (butter and margarine). A firm that provides two or more vertically related products is said to be *vertically integrated*. (For a picture of how the electricity industry is vertically integrated, see figure on page 84.)

In any industry, firms have to decide how much vertical integration is appropriate, both in terms of *scope*—how many production stages are covered—and *degree*—how detailed and complete is the control of one production stage by another. At its most basic level, the decision can be framed as a "make/buy" choice: does the firm make the product or perform the service it needs, or does it purchase that product or service from a separate provider in the market? To appreciate the virtues of vertical integration in electricity and to discern whether those virtues outweigh potential vices, it helps to understand the subtle and interrelated factors on which the make/buy choice turns. The most

Three functions
of the
electricity
industry
organized
into a single,
vertically
integrated
company or
into three
separate
companies

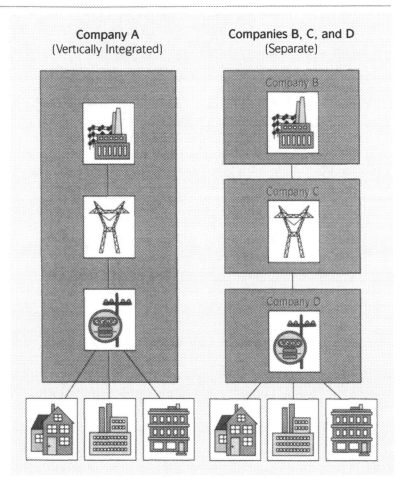

prominent considerations, regardless of the industry, behind decisions to vertically integrate are transaction costs, economies of scope, improving coordination, and hedging risk.

Transaction Costs: Using Markets Is Not Cost-Free

Competitive markets, in theory, are perfectly effective means for buyers to obtain the most suitable goods at the least cost and for sellers to find the buyers willing to pay the highest price for their products and services. For vertical integration to be beneficial, there must be substantial costs to using markets to find sellers of the goods and services one wants and buyers for one's products. Because what one does in markets is transact business, the costs of using markets to reach mutual agreements between buyers and sellers have come to be known as *transaction costs*. If transaction costs were negligible, there would be no need for vertical integration. However, virtually every aspect of a market transaction—from discovering one's needs to identifying relevant providers, to verifying product characteristics, to establishing and negotiating price, to ensuring timely delivery—may have substantial associated costs. A business may be able to avoid those transaction costs by integrating upstream into the markets for inputs or downstream into the markets where its outputs are used or retailed. For instance, an oil refinery that makes gasoline may choose to acquire a firm upstream that supplies crude oil or a firm downstream that delivers and sells gasoline to final customers.

Most of the proposals for restructuring the electricity industry—from open-access rules, through fully separate subsidiaries, to explicit divestiture of generation from transmission—call for some lessening of the industry's traditional vertical integration. Would these proposals lead to a substantial increase in transaction costs? Our impression at this time is that they probably would not. Transaction costs are more likely to be high when the product one wishes to buy or sell is highly *differentiated*—that is, with many variable characteristics (like, say, hiring a contractor to remodel your home)—rather than *standardized*—that is, with a few, easily compared characteristics (like, say, buying paper clips). If electricity is in large degree a standardized product, as some industry observers believe, then the transaction costs would be low and the efficiency gains of vertical integration would be small. But if power characteristics vary considerably among generators, or if restructuring leads to differentiated electricity services tailored to the needs of particular customers, then creating a situation where, for instance, generation and transmission companies have to deal with each other at arm's length could have substantial associated costs.

Economies of Scope

Vertical integration is frequently justified on the grounds that it is simply less expensive to produce two products at two adjacent stages in the production chain in one place than it is to produce the same two products by separate companies. The idea that joint production of *multiple* goods or services is cheaper than separate production of each of them is referred to as *economies of scope*. (For more on economies of scope and economies of scale, see box on page 86.)

In the electricity industry, economies of scope might arise where two of the functions share the same physical facilities—such as where generators interface with the transmission grids or the grids interface with end users. So far, the opportunities to share costs of these physical facilities in the electricity industry do not appear to be great. Prior, ongoing, and future research comparing the performance of vertically integrated utilities with that of nonintegrated distribution systems and generation companies should shed additional light on this question.

What is economy of scale? of scope?

Economy of scale describes situations where quantity counts: where the average cost of producing an item falls as the number produced increases. In other words, the more of an item that a company makes, the less each item costs to make on average. When there are economies of scale, it is less expensive for one company to produce a given amount of a single product than to have that amount produced by separate companies.

Economy of scope describes situations where range counts: where the joint production of multiple items is cheaper than separate production of each of them. When there are economies of scope, it is less expensive for one company to produce two products at two adjacent stages in the production chain than it is for two separate companies to produce the same two products.

Improving Coordination

A common reason for a company to vertically integrate—or to enter into vertical contracts with firms at different stages of production—is so that each actor along the vertical chain is encouraged to take the interests of everyone in the chain into account. For instance, an automobile company builds its own engines, rather than buying them from an outside provider, because such vertical integration allows it to coordinate engine design with the construction of the body and other components of the car. Even when a company does not vertically integrate itself, vertical contracts help align incentives to market a product; for instance, an exclusive license between a manufacturer and a retailer may encourage the latter to promote the former's products, without fear that new business would be lost to other retailers who do not bother to advertise.

In the electricity industry, important cost savings could possibly be achieved through integrated coordination of dispatch, load management, reliability maintenance, and other services associated with power delivery. As with economies of scope, the extent of these cost savings is a crucial subject warranting additional research.

Hedging Risk

Vertical integration allows a company to hedge its bets by operating in such a way that, typically, bad times in some markets will be offset by good times in the others. For example, for an integrated electric utility, a decline in electricity prices resulting from increased competition could reduce profits among generators, but this might also increase demand for transmission services, increasing profits in the transmission sector (assuming, of course, that transmission regulators do not act rapidly to cut those profits by reducing transmission rates). A utility could maintain the same average return on its investments over time but reduce its exposure to risk by operating in both the generation and the transmission markets.

Vertical integration, however, is not the only means of achieving this result. Investors can hedge risk on their own by owning shares of independent firms operating at different stages in the production chain. Consequently, all else being equal, hedging risk is not a compelling reason to maintain vertical integration in the electricity industry.

Vertical Integration Where Markets Are Not Competitive

The potential benefits of vertical integration that we just enumerated apply generally throughout the economy and, for that reason, make vertical integration a virtually universal phenomenon. The degree of vertical integration varies across industries, with electric utilities that own their own coal mines—as some do—being nearly as integrated as they can be. In markets where competition drives prices close to cost, the degree of vertical integration is rarely a policy concern. If one firm's vertical integration is too little or too much, it will find out soon enough. When a company fails to reduce costs below what it could achieve by relying on markets instead, the company will lose market share and profits to less integrated, more flexible, lower-cost competitors.

If markets, however, are not competitive—especially if some of those uncompetitive markets are regulated, as is likely to be the case in electricity for some time yet—the merits of vertical integration may be harder to identify. In some cases, a lack of competition could *increase* the net benefits of vertical integration. Imagine a transmission company and a local distributor, each with some ability to raise price above costs (for example, because regulation was either ineffective or absent). When each raises its price to increase its own profits, it reduces demand for electricity or it raises costs for the other, thus reducing the other's profits. If the transmission company and the distributor coordinated their pricing, they would take these negative effects on each other into account and set their prices below what they would each have set independently. The result is both increased profits for the transmission and distribution companies *and* lower prices for electricity consumers.

Vertical Integration under Regulation

Sometimes, in uncompetitive markets, vertical integration works in the other direction, leading to less competition and higher prices. These outcomes are particularly likely when vertical integration crosses the boundary between regulated and unregulated markets—as we may have in the future, say, if individual companies provide both regulated transmission and unregulated generation.

The electricity industry is not unique in posing problems by spanning regulated and unregulated markets. During the late 1970s, policymakers were concerned about the possible anticompetitive consequences of vertical integration by competitive oil and gas shippers into the regulated pipeline business. Later on, the view that regulated local telephone companies should not operate in other competitive communications markets, such as long-distance service, led to the divestiture in 1984 by AT&T of its local telephone companies and to restrictions on the reentry by those companies into competitive markets. The controversy that ensued from these restrictions—continual court and regulatory challenges, with prospects for resolution offered only recently through the Telecommunications Act of 1996—foreshadows controversies over the need for parallel policies when it comes to restructuring the electricity industry.

Legal and political attention to vertical integration in regulated contexts arises because regulation typically forces a company to set its prices at a level below what would maximize its profits. Consequently, the company has a financial incentive to look for ways to evade the regulatory constraints. The history of the electricity, pipeline, and telephone industries identifies several ways regulated firms can exploit vertical integration to exercise the market power that regulation is designed to control. These tactics include self-dealing, undersizing, discrimination, and cross-subsidization.

Self-Dealing

Self-dealing is the practice of selling to oneself, as shown in the following example. Suppose an electric utility is vertically integrated upstream into the production of one of its inputs, for example, coal for its generators. If the regulator allows the utility to raise the rates for its regulated services when it pays higher prices for inputs, then the utility might sell coal to itself at artificially inflated prices. In this case, the rates go up along with the utility's "costs," so the balance sheet for the utility side of the business remains even. The profits taken from the utility's electricity customers show up, instead, on the coal side of the business.

As we mentioned above (and discussed in Chapter 2), Congress passed the Public Utility Holding Company Act in 1935 in part to address this kind of inappropriate self-dealing. In the above example, self-dealing is not likely to be a severe problem, because products such as coal are sufficiently generic to make it easy for regulators to compare the prices that a vertically integrated utility charges to itself with those in the market. To the extent that the corporate boundary affected by restructuring is that between generation and transmission, we would not expect self-dealing to be much of a problem. Because transmission companies do not use much electricity, except perhaps to maintain load balances (as described in Chapter 4), they have little to gain by selling themselves electricity at inflated prices.

Undersizing

The *undersizing* story may apply when a regulated firm is owned by a collection of upstream unregulated firms that use it. These firms then would have an incentive to limit the capacity of the regulated firm, thus reducing output and raising product prices to downstream buyers. The concept of undersizing was originally developed to explain potential anticompetitive behavior regarding oil company ownership of regulated oil pipelines. Its relevance in the current context is that a consortium of unregulated generators owning a regulated transmission company could limit the transmission grid's capacity to deliver electricity. The reduction in power supplies resulting from this undersizing would create a premium for electricity in downstream markets, from which the generators could profit. Thus, power producers would have an incentive to own the transmission system and then undersize it in order to get around the regulation of transmission prices and to capture the monopoly profits themselves. Moreover, absent regulation, a separately owned transmission company could capture these monopoly profits directly through high shipping charges.

If vertical integration leads to significant problems of undersizing, antitrust enforcement may prevent collusion to keep the grids small. Regulatory authority to order expansions, such as that provided to the Federal Energy Regulatory Commission under the Energy Policy Act of 1992, also may help prevent anticompetitive limitations on the capacity of the transmission grid.

Industry observers continue to debate whether either of these methods could be more timely and effective than divestiture in addressing undersizing.

Discrimination

In essence, *discrimination* occurs when efficient access to a regulated company's services is limited to those who also purchase related products and services from its affiliates. For instance, suppose that despite legal restrictions a regulated transmission firm could grant preferential treatment to its own generators. In such a scenario, the transmission company then could charge a premium for its electricity, thereby profiting from its control over the regulated service. Without the market power endowed by the regulated monopoly in transmission, the utility could not charge the premium, since its competitors in the generation business could obtain transmission elsewhere and would not have to put up with the discrimination.

> *Because of concerns over abuses (such as under-sizing and discrimination) that may arise from vertical integration, some analysts and regulators argue for the functional separation, if not outright divestiture, of generation from transmission and distribution.*

In the telecommunications industry, the primary basis for the federal antitrust case against AT&T was that, in long-distance markets and equipment sales, AT&T discriminated against its competitors, who needed to connect their facilities and products to the regulated local telephone companies that AT&T then owned. As with undersizing, perhaps nondiscrimination rules will suffice, or perhaps more drastic structural approaches—such as corporate separation (with an independent transmission system operator) or divestiture—will be necessary.

Cross-Subsidization

If regulators set a regulated utility's prices on the basis of costs, the utility has an incentive to inflate those costs and secure higher prices. One tactic begins with diversifying into an unregulated business, to enable the regulated firm to charge some of the costs incurred to provide the unregulated services to the customers of its regulated service; this is called *cross-subsidization*. A hypothetical example from electricity would be if a regulated transmission company designated some costs of generating electricity as "transmission costs" and then used those "costs" to justify an increase in rates. As with self-dealing, the profits would show up on the books of another part of the business—in this example, the generation side, which has been cross-subsidized by revenues generated on the regulated transmission side of the business.

Cross-subsidization is more likely when state or federal regulators find it difficult to verify whether reported costs were incurred to provide the regulated or unregulated services. Paradoxically, it is also more likely, all else being equal, the more closely regulators tie prices to costs. If regulation is lax, a regulated firm can exploit more of its market power directly, without having to resort to diversification just to evade those regulatory constraints.

Key Restructuring Questions

Responding either explicitly or implicitly to the above concerns about the abuses that could accompany vertical integration, many policy analysts and regulators today are considering the need for functional (and perhaps actual) separation of the competitive generation market from regulated transmission and distribution services. It is theoretically possible for regulators to deal with potential abuses by controlling reported self-dealing prices, prohibiting capacity restrictions, imposing antidiscrimination rules, and policing cost reports to prevent cross-subsidization. Unfortunately, the history of regulated industries (primarily of the telephone industry, but also of electricity) inspires something less than complete confidence that merely instituting such rules will lead to effective monitoring and timely enforcement. If rules alone are not enough, then policymakers are left with difficult questions, such as the ones that follow.

How much will future regulation discourage anticompetitive conduct?

Both self-dealing and cross-subsidization require that the regulators base the regulated firm's rates closely on its reported costs. Under price-cap regulation, as described in Chapter 4, rates would be divorced from costs, eliminating incentives a utility might have to achieve profits by fabricating cost data, at least after the price caps have been set. As we have noted already, however, regulators are likely to find it difficult to fulfill public commitments to keep prices separated from costs, leaving some incentive for a utility to attempt to profit from self-dealing or cross-subsidization. Moreover, price-cap regulation does not address the possibility that a vertically integrated utility could profit by discriminating or undersizing its regulated transmission facilities.

Would a regulated transmission or distribution grid that is integrated with generation have the opportunity or ability to engage in anticompetitive activities?

On the financial side, the history of the Public Utility Holding Company Act suggests that, prior to its passage, some utilities engaged in advantageous self-dealing by using inflated charges for internally supplied services to induce regulators to raise prices to consumers. When it comes to the current and future marketplace, however, there are just too many unknowns at the moment to confidently answer this question. For example, are there significant costs associated with unregulated services that could be allocated to transmission and distribution rates? If the responsibility for covering the capital risk of unregulated operations were shifted to the grid's regulated operations, then a cross-subsidy effectively could be generated from the latter side of the business to the former. Diversification by utilities into appliance markets and demand-side management services raises similar questions. Research to discover whether utilities put more or less effort into these activities when they are under price-cap regulation, compared to when they are under conventional rate-of-return regulation (as discussed in Chapter 4), might provide a useful test of whether costs can be misallocated to the regulated side of a partially competitive, partially regulated utility.

What opportunities to discriminate are likely to exist?

If a vertically integrated transmission or distribution grid owner becomes actively involved in dispatch of electricity, it may have an incentive to ensure that its own generators are preferentially dispatched. For this reason, most competition proposals declare that the Poolco or system operator should be an independent entity. A more subtle, but perhaps no less effective, tactic for discrimination would be if a transmission company more promptly supplied and better maintained the lines it uses to connect to its own generators than the lines it uses to connect to competing generators. This could increase the cost and decrease the reliability of power coming from the competing generators, creating an inappropriate market advantage for the generators affiliated with the transmission company.

How much must be done to head off discrimination and cross-subsidization?

Astute and effective regulators can limit the severity of problems created by discrimination and cross-subsidization. Each of these tactics relies on a somewhat paradoxical premise. For discrimination to be profitable for a utility, electricity prices have to rise in response to the effects of discrimination, and yet state and federal regulators have to be unable to infer discrimination from these effects. Effective cross-subsidization requires that regulators are diligent enough to connect transmission or distribution prices to reported costs, but insufficiently diligent or able to determine whether those costs were indeed associated with regulated transmission or distribution services. Neither paradox implies that anticompetitive conduct is impossible, but policymakers should be circumspect about requiring more stringent means of preventing it, such as the entry restrictions that were imposed in the telephone industry in 1984 to prevent vertical problems from recurring following the AT&T divestiture.

Conclusion

Beyond the issues just cited, many more will require careful theoretical and empirical examination as regulators and the electricity industry go about restructuring. For example, municipal distribution companies may wish to vertically integrate, but as government-owned companies, their particular incentives and goals may make the effects of such integration difficult to predict and evaluate. As another example, jurisdictional issues, particularly federal preemption of state rules affecting the diversification of utilities, could be evaluated according to whether the harm, if any, of diversification can be exported across state lines.

Chapter 6

Paying for the Past before Entering the Future
Coping with Stranded Costs

If the intensity of policy debate were to be measured by the number of words in the trade press or by the size of the economic consequences that the debaters claim, then no other restructuring issue stacks up against stranded costs for well-inked pages and dollar signs with lots of zeros attached. In this chapter, we attempt to sift through the issues surrounding "stranded costs," "stranded assets," or, as some propose calling them, "embedded costs exceeding market prices."

By any name—we use *stranded costs*—those terms refer to the prospect that, as the electricity industry becomes more competitive, some utilities may not be able to earn enough to recover the costs of their prior investments in power plants, of their long-term contracts with

95

PURPA-designated qualifying facilities, and of other regulatory obligations. The amounts at stake are not insignificant. According to an article in Public Utilities Fortnightly, the electricity industry is facing stranded-cost losses that range from $10 billion to $200 billion, depending largely on what assumptions are used regarding the breadth and depth of competition. (See figure on page 100.) The political and economic success of efforts to expand competition in the electricity industry are likely to depend very much on how policymakers deal with stranded costs.

As state public utility commissions and the Federal Energy Regulatory Commission grapple with the stranded-cost problem, they will need the answers to several important questions.

- How big is the stranded-cost problem in the electricity industry?
- Was it caused by poor planning by utilities, misguided public policy, or unforeseeable changes in circumstances?
- Is there now, and has there always been, a so-called "regulatory compact" obligating federal and state governments to guarantee utilities at least a reasonable opportunity to recover costs, if not to guarantee full cost recovery?
- Have utilities exploited their end of that compact by building plants without proper regard to need or cost? What could or should regulators do now to mitigate the problem?
- Should the utilities' shareholders—including workers, through their pensions, as well as capital investors—be forced to swallow losses in a competitive environment for electricity, or should the utilities' customers be obligated to compensate them?
- If customers should pay, which customers should pay, and how and when should electricity prices be increased over the prices that might otherwise prevail in a more competitive electricity market?

We cannot resolve these questions in this chapter, but we can suggest where to look for answers and what form those answers are likely to take. First, we examine stranded costs throughout the economy to gain some insight as to why the electricity industry might warrant an exception to the rule that firms get and deserve no public protection against the effects of competition and innovation. Then, we consider the economic principles of regulation that support the idea of accelerating cost recovery, should regulators decide to open retail and wholesale electricity markets. Next, we review the economics of contracts to identify factors that determine whether expanding competition in the electricity industry represents a breach of a so-called "regulatory compact" between utilities and the public. Finally, because a decision that the public should bear those risks should depend upon the effects of practical solutions, we look at some of the proposals for recovering stranded costs that are now circulating throughout the industry.

Stranded Costs throughout the Economy

Innovation, product development, changes in consumer taste, and many other social, political, and economic factors frequently render obsolete factories, equipment, and skills that previously commanded high prices. Examples abound. Thanks to personal computers, keypunch machines are a thing of the past. Audiocassettes wiped out eight-track cartridges, and compact discs have eliminated much of the profit from vinyl sound recordings. Automated "just-in-time" factories with robotic assembly lines led to the abandonment of old-style factories, which took lots of blue-collar jobs when they went. Such obsolescence is part of the process of "creative destruction," to which economist Joseph Schumpeter attributed so much of the impetus for economic growth. As innovations or changes in demand take place, factories are left for the bulldozer; machinery is sold for scrap value; and workers face wage reductions, layoffs, and the need for retraining.

Innovation, product development, and changes in consumer taste can make factories, equipment, and skills obsolete. Whether electric utilities could have responded to potential changes in technology and regulation is at the heart of the debate.

If competition in the electricity industry becomes more widespread, electric utilities and their regulators encounter the possibility that many of the assets used to produce and purchase electricity will suffer a loss in value due to just this obsolescence. In industry parlance, the *costs* of these assets would become *stranded*—that is, money will have been invested that cannot be recovered. The conjectured scenario goes like this: the advent of competition and the entry of new, low-cost generators cause electricity prices to fall. A utility that had built an expensive nuclear plant; that operated older, high-cost generators; or that had signed a long-term purchase contract at high rates might be unable to sell its electricity at prices high enough to fully recover the costs of these plants or contracts. The unrecovered costs are left stranded. As the generation side of the industry is most likely to face growing competition, and as it has been the focus of both expensive investments in nuclear plants and high-cost contracts for power from alternative sources, it is the segment in which the stranded-cost problem is most pronounced.

Could Electric Utilities Be a Special Case?

Since businesses throughout the economy regularly absorb losses brought about by changes in technology or competition—cushioned only by write-offs to reduce their corporate income taxes—it might seem that electric power generators should not be treated any differently. Two factors may make electric power generation worthy of special concern as competition grows.

First, markets allow typical businesses the flexibility to respond to expected changes in their competitive and economic environments. In regulated markets, however, prices are set administratively, often according to prescribed accounting rules and depreciation schedules. Prices are not set by market forces responding to changes in risk. To see how this might affect power companies, suppose that the useful lifetime of a particular type of power plant is expected to decrease because technological change will render it obsolete. If utilities were operating in a competitive electricity market, they would not install or replace a generator until prices rose to compensate them for the risk of obsolescence. Under regulation, rates might not increase sufficiently to reflect that risk.

The second reason electricity generation may merit special consideration is that policy choices themselves may cause costs to be stranded. While technological and economic pressures play a role, explicit decisions by state and federal regulators—to discourage energy demand, to encourage renewable forms of electricity production, and to permit entry and competition—also may be responsible for stranded costs.

In and of itself, this second factor is not sufficient to warrant special policies to protect electric utilities. Government programs routinely result in stranding costs throughout the economy, as when building an interstate highway destroys roadside stores formerly dependent on through-traffic or when publicly funded research leads to innovations that make some products obsolete. In the past, some stranding has been a consequence of public decisions to permit entry and competition. For example, when regulators allowed MCI, Sprint, and other long-distance telephone companies to use microwave and fiber-optic technology to transmit voice and data, they accelerated the decline in the profitability of AT&T's old copper-based intercity network. As another example of policy-related stranding, a free-trade agreement could increase competition in a previously protected industry. It is important to remember, however, that investors in telephone networks, roadside stores, or other enterprises would require higher returns before they invest to the extent that they anticipate greater risks of public policies stranding their assets—just as those investors would incorporate into their investment decisions the risk of nongovernment-related influences affecting profits.

Could Regulation Be To Blame for Stranded Costs?

In the electricity industry, regulation may not only have precluded adaptation to policy changes; it also may have made electric utilities more vulnerable to risks arising from policy changes. Those arguing that utilities should be allowed to recover stranded costs contend that many of the generation facilities and expensive energy contracts that could be stranded under competition were not incurred by utilities taking business risks but, rather, were the result of regulatory and legislative mandates. Were it not for orders from public utility commissions, these advocates say, utilities would not have built facilities to meet regulation-mandated obligations to serve the entire public, nor would they have signed long-term contracts to purchase energy from expensive alternative sources. Other proponents of stranded-cost recovery maintain that regulatory approval of these contracts and construction projects suffices to ensure that the utilities should be able to recover the cost of these investments.

These possibilities set out the basic premises for an analysis of the stranded-cost problem. As entry and competition grow, regulators may need to allow rate increases that take into account the possibility that electric utilities may soon face changes in demand in some or all of their markets for electricity. As deregulation continues, policymakers will have to decide whether and how much to compensate utilities for costs left stranded once passing them on to consumers is no longer an option.

Economic Principles of Regulation: How Big Could the Stranded Costs Be?

As we noted at the beginning of this chapter, estimates of stranded costs spread across an enormous range, largely because it is so difficult to forecast how much electricity costs and prices will be reduced by competition and to predict the volume of entry and exit into electricity markets. (See figure on page 100.) In some situations, rates may well rise for some users and fall for others, as the competitive process squeezes subsidies out of the rate structure. On one hand, removal of regulation should give utilities a greater incentive to control costs, and this should lead to lower prices, making it harder for a utility to recover its stranded

Wide
variations in
estimates of
stranded
costs

For policymakers and industry experts alike, the stranded-cost problem is exacerbated not only by its magnitude, but also by the range of estimates available, shown in the accompanying chart. The estimates run from as low as $20 billion (according to the American Public Power Association) to upwards of $200 billion (according to Niagara Mohawk Power Corporation), measured against industry equity of approximately $180 billion.

The range of estimates can vary so widely because of uncertainty in many factors: the future price of electricity; the length of time the predicted prices will last; the discount rate by which future losses are translated into present losses; the scope of competition, as in whether it will reach residential and commercial buyers as well as industrial customers; and the degree to which these losses can be written off against other taxable income. For example, Hirst and Baker offer two different estimates. In estimate #1, competitive prices are quite low, but only industrial customers benefit from them; in estimate #2, all customers benefit from competition, but the predicted price of electricity is not as low as in #1.

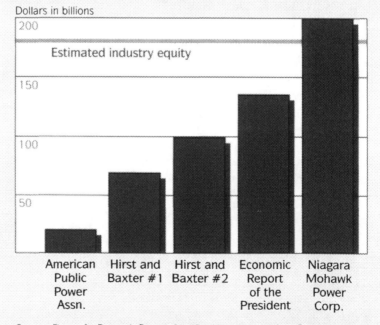

Dollars in billions

Sources: Figures for Economic Report of the President are taken from Economic Report of the President, Feburary 1996; other figures are from Eric Hirst and Lester Baxter, "How Stranded Will Electric Utilities Be?" Public Utilities Fortnightly, February 15, 1995.

costs. On the other hand, removing regulation could increase a utility's profit margins and, thus, its ability to recover preexisting costs. Some utility assets may be worth more in an increasingly competitive environment, and these gains should be counted against any stranded-cost liability.

In addition, estimates of stranded costs should reflect the possibility that overall prices may not fall by much, at least in the short run. In regions of the country with little excess capacity in electricity generation, prices are not likely to fall significantly as competition grows, especially if regulators had set prices close to average cost. While new plants may be more efficient than many older generators or than many of the qualifying facilities under contract to utilities, new power suppliers will not enter unless they expect that electricity prices will be sufficiently high to cover all of the costs of building these new generators. For incumbent utilities to recover their stranded costs, however, they need only raise enough revenue to cover existing contract and generation costs, which have been recovered at least partially already.

Moreover, unless new policies are enacted, regulators may not allow substantial generation and power contract costs to be left stranded. After all, under some theories about the regulatory process, regulators will tend to be responsive to the desires of the utilities. Nevertheless, regulators can make themselves unpopular with ratepayers by burdening them with the costs of recovering investments that are no longer "used and useful." In addition, regulators may be unwilling to raise rates prior to the onset of full competition if the public suspects that doing so would be a vehicle for anticompetitive cross-subsidization—for example, if utilities were allowed to raise rates today to facilitate future below-cost pricing when they face competition from new providers.

If stranded costs are as sizable as many industry observers warn, the economics of regulation suggests that electricity rates should be permitted to rise in the short term in anticipation of expanded competition, allowing for some additional cost recovery. For example, regulators could begin now to allow utilities to accelerate the rate at which they depreciate their investments. Of course, if the move to competition is rapid, increasing rates within the current regulatory structure may not be possible. Widespread and quick entry into most or all of a utility's markets—wholesale and retail—would limit the ability of accelerated depreciation alone to raise sufficient funds to cover the stranded costs. Moreover, political resistance to rate increases may invite regulators to strand the regulated utilities' assets, leaving the utilities' shareholders with the losses.

Economics of Contracts:
What Was the Regulatory Compact with Electric Utilities?

As regulated industries, electric utilities are in some respects parties to a contract with the government. Electric utility investors constructed generation, transmission, and distribution facilities—and, more recently, they entered into long-term contracts with nonutility generators—to provide electricity to the public. In exchange, the government approved these investments and established the rules for the utilities' operation; these rules include one that prices shall be set sufficient for the utilities to have the opportunity to earn a "just and reasonable" return, as specified in 1944 by the Supreme Court in *Federal Power Commission v. Hope Natural Gas Co.* If public policies to permit competition in electricity were to be interpreted as a breach of a regulatory contract, then the government would be as obligated as any other breaching party to compensate the utilities for the losses attributable to this breach. The question of the moment is whether this is a fair interpretation to place on relations between electric utilities and the government.

Is There a Regulatory Compact?

Protection of asset value is of the utmost importance in regulated industries. Regulated firms, including electric utilities, typically must make large, fixed, irreversible, and long-lived investments in order to provide their services. Once a utility sinks those costs, it is vulnerable to a regulatory change of heart. A regulator could refuse to allow the firm to receive more than what it would take to cover the bare costs of operation, so the regulated firm has no chance to recover its sunk investment, much less to earn a return on it. Without protection against this potential exploitation, investors would not supply the capital necessary to provide the regulated services. For regulated industries in general, unless public ownership of the sunk costs is economical, the primary protection is in Supreme Court doctrines guaranteeing investors at least the opportunity to earn a "just and reasonable" rate of return.

Identifying the relationship between a regulated firm and the state as a contract still only frames the question of who pays for stranded costs; it does not answer it. Explicit, unambiguous provisions in regulations, statutes, or constitutions could provide a definitive interpretation as to who is responsible for stranded costs. Unfortunately, neither the implicit compacts between utilities and regulators nor explicit rules set by regulators address directly the question of stranded-cost recovery. Lacking a customary agreement or specific guidelines, the question of stranded costs could be addressed by asking how the parties would have resolved this problem had they had the foresight and ability to agree on how to treat stranded costs were the contingency to arise.

Suppose, then, we try to determine how an ideal regulatory contract between federal and state governments and the electric utilities would have dealt with stranded costs. We quickly come up against the problem of *moral hazard,* a term economists use to refer to exploitation of commitments. When two parties, such as an electric utility and a regulator, have made significant, irreversible, and specific commitments to each other, each may have the ability to take advantage of the other. In a perfect world, there would be no moral hazard—all behavior could be specified in advance, monitored, and rewarded or penalized as needed. Unfortunately, the costs of specifying contingencies, monitoring behavior, and enforcing penalties all raise the possibility of moral hazard coming into effect.

> *Once a utility sinks money into large, long-lived investments, it is vulnerable to a regulatory change of heart. Without protection against losing sunk costs, investors would not supply the capital needed to provide the regulated services.*

Moral hazard can apply in either direction in the context of stranded costs. On one side, knowing that a utility cannot relocate a facility once it has been built, a regulator could renege on its promise to allow utility investors to recover their investments in power plants or transmission lines. Apart from considerations of fairness, such reneging would reduce the credibility of future regulatory promises to investors that they would be able to recover their investments in regulated firms. On the other side, if a utility believed that a regulator was obliged to cover all its costs, then the utility would lack an incentive to avoid investments that were not economically justifiable. This is just what some critics of the utility industry have charged—that utilities should have been more cautious in overseeing the cost of constructing generation capacity and contracting for power; they argue that utilities were not sufficiently cautious because the utilities assumed that regulators automatically would pass the costs along to ratepayers.

Who Bore the Risk of Stranded Costs?

From this discussion, it should be clear that the controversy over stranded-cost recovery is not simply about measurement, but also about the assignment of responsibility for incurring costs and contracts in the first place. Properly assigning responsibility for stranded costs, however, requires an understanding of who was in the best position to bear the risk of subsequent competition at the time the investment decisions were made. At first blush, the government seems to be the best risk-bearer. Competition depends on public decisions to reform regulation and remove restrictions on entry. If so, the government presumably was in a better position to predict its own future policies. Accordingly, the government—that is, the public at large—should bear the burden of recovering the costs left stranded by those policies.

Three possibilities, however, suggest that the utilities might be the best risk-bearer. First, regulators already may have compensated utility shareholders for that risk of stranded costs, if the investors were allowed to earn a rate of return reflecting the chance that future competition might preclude full cost recovery. Second, to the extent that expansion of competition has been driven by changes in technology and demand, the utilities may have been in a better position to forecast and insure against competitive risk than their regulators. Third, many of the proposals to expand competition in electricity markets arose at the federal level, while decisions on rates of return and prudence of investment are made by state public utility commissions. State regulators may not have had an advantage over utilities in predicting what the federal government would do.

In looking at stranded costs from the perspective of the economics of contract interpretation, the crucial issue is not so much one of fairness in assigning stranded costs as in the long-term consequences of how that assignment is made. A rule that looks generous from the utilities' standpoint will send a signal to regulated firms that they need not worry about the prospect of future competition when planning and advocating investments. If too little recovery is permitted to the utilities, however, then investors will be discouraged from providing the capital necessary to support regulated services in the future. Beyond the economic issues is a potential political imperative. Some industry observers argue that regulatory reform and the acceleration of competition in the electricity industry cannot proceed without the cooperation of utilities and that stranded-cost recovery will be necessary to secure their support.

Recovering the Costs:
Practical Proposals for Handling Stranded Costs

Even without current pressures on federal and state budgets, the reality is that government is not an independent source of wealth that claimants can tap. If economic rationales, political imperatives, or judicial decisions bring about a situation where substantial stranded costs must be recovered, the important issue facing policymakers will be to determine how the necessary payments will be made. Eventually, those payments *will* be paid by the customers to the shareholders and creditors of the utilities.

Regulators, economists, and industry experts have offered numerous proposals for recovering the cost of the assets that would be stranded by competition. Among these proposals are the ones that follow.

Contract Renegotiation

In Chapter 4, we discussed the Federal Energy Regulatory Commission's (FERC) Order 888, which was issued in April 1996, regarding open access to transmission lines. In the order, FERC suggested that a utility could renegotiate contracts with its wholesale customers to extract compensation for stranded costs, unless those contracts already included some explicit stranded-cost payment provision. In effect, Order 888 allows a utility to reap whatever it can from its customers. The customers with competitive alternatives could turn elsewhere, but those who do not could be quite vulnerable. The most vulnerable customers are likely to be those who had made other capital commitments to use power from the utility, based on the expectation that electricity rates would be regulated. Some industry observers question, however, whether utilities could benefit from renegotiation. Regulatory obligations to serve all customers may limit the ability of utilities to negotiate higher payments that would help recover stranded costs.

Transmission Surcharges

As we noted in Chapter 5, utilities that own transmission facilities are likely to retain market power for some time to come. Were this not so, it would not be necessary to have rulemakings on open access, because competition among transmission systems would ensure that those willing to pay the cost of transmission would be served. To the extent that utilities retain market power over transmission, however, regulators could allow them to add a surcharge on all transmission services to contribute to recovery of stranded costs associated with generation. Unlike the renegotiation proposals, the transmission rates would remain regulated, but at a higher level than they would otherwise be.

As a rule, it is more efficient to charge higher surcharges to those customers whose demand for transmission services is the least sensitive to price. This minimizes the economic losses associated with any reductions in electricity use resulting from these higher surcharges. However, applying this rule may not be politically feasible, nor will it necessarily allocate the burden of paying for stranded costs to those customers responsible for incurring the stranded costs—primarily industrial customers, who have greater ability to turn to alternative means (including cogeneration) for acquiring power.

Compensatory "Exit Fees"

When a transmission surcharge plan is being considered, it must somehow include a strategy for dealing with customers that strand a utility's generators or power contracts by seeking electricity from other suppliers. These customers, however, frequently still require that the utility transmit power from that other supplier to the customers' premises. Yet another proposal for stranded-cost recovery suggests that these customers pay the utility an *exit fee* on top of its transmission costs, to cover the generation costs that their departure strands. In Order 888 regarding open access to the transmission grid, FERC suggested that the customer could pay the utility with stranded costs an amount calculated by subtracting the "competitive market value" of the electricity from "the revenues the customer would have paid the utility" had it not left. Other proposals base exit fees on the contributions to stranded-cost recovery that would otherwise be lost when the customer switches to another power supplier.

Government is not an independent source of wealth that claimants can tap. If substantial stranded costs must be recovered, policymakers must determine how the payments will be made.

Exit fee proposals have some intuitive appeal. Compensating the incumbent utility for its lost profits means that new generating companies will survive only if their costs are less than the incumbent utility's incremental cost of serving the customer. Such exit fees thus would prevent entry by inefficient generators who are only competitive because they do not need to cover stranded costs, as the utility does. In the long run, however, exit fees may stratify the current industry structure, preventing the benefits of competition from accruing to end users and the economy as a whole. Under FERC's Order 888, a customer would not benefit from switching to a cheaper supplier since the exit fee it would have to pay its old supplier would be equal to the savings it would otherwise enjoy. Thus, the exit-fee rule would discourage entry by and competition from new power suppliers.

Accelerated Depreciation

Until full competition becomes a reality, regulators may be able to allow some stranded-cost recovery by increasing the capital recovery allowance used to compute rates, in effect accelerating depreciation. Merrill Lynch has suggested increasing the capital allowance by changing the transmission rate base from depreciated book value to the much higher replacement cost figure. Accelerated depreciation in various guises can be a justifiable response to the onset of restructuring, so these proposals may promote efficiency as well as reflect other pragmatic concerns associated with cost recovery. As noted earlier, the usefulness of this remedy is limited if competition comes quickly and prevents utilities from charging the higher rates that accelerated depreciation would allow.

Auctioning Off Transmission Facilities

Some industry observers have suggested allowing utilities to auction off the transmission lines that connect power plants to local distribution systems, using the proceeds to cover stranded generation costs. Unless the winner of the auction can charge rates in excess of those based upon the undepreciated book value of the transmission plant, however, the auction would raise only enough funds to recover transmission plant investment costs. There would be nothing left over for stranded-cost recovery. As a consequence, the ability of an auction to cover stranded costs depends on whether regulators allow suppliers of transmission capacity to charge rates above costs. If the regulators do so, however, the auction makes little difference; the utility could generate the stranded-cost recovery revenues through those higher fees. Therefore, the effectiveness of an auction proposal essentially depends upon the effectiveness of these regulatory pricing policies; it is not an independent remedy for the stranded-cost problem.

Making Stranded-Cost Recovery Depend upon Environmental Improvements

Some environmental advocates have proposed to permit utilities to recover stranded costs in return either for bringing existing plants into compliance with stringent emissions standards applicable to new plants—such as the Clean Air Act New Source Pollution Standards (NSPS)—or for purchasing emissions permits if those plants fail to meet those standards. From an environmental policy perspective, this approach has the advantage that it helps to counter a bias that arises because the cost of extensive pollution controls makes constructing and operating new plants more expensive than operating existing plants that are subject to somewhat less stringent regulation. However, the appropriate emissions cap is one that meets environmental objectives, which need not be achieved when all sources are meeting NSPS. In exchange for the opportunity to recover stranded costs, utilities might be *too* willing to accept NSPS for older generators, in the sense that the social benefits from reduced pollution may be less than the economic cost of abatement. The most critical problem with this proposal is that it mixes environmental issues with stranded-cost concerns, when both are separate problems meriting separate, focused remedies. (We consider the environmental implications of restructuring in Chapter 7.)

Conclusion

As the prospect of competition looms larger, the pressures to answer the stranded-cost questions—how large, who should bear the burden, and what might be done about it—will become even greater. Policymakers may find it impossible to bring the benefits of competition to electricity markets unless they can address effectively the stranded-cost concerns of utilities, their competitors, and their customers. Deciding how much stranded-cost recovery is necessary should begin with a careful analysis of the extent to which the utilities or the public at large were better placed to cope with the risks posed by potential growth of competition in generation. Regardless of the motivation for stranded-cost recovery, the method of compensation to choose should depend at least in part on its economic effects. These effects include the short- and long-run effects on electricity prices and, as we discussed in Chapter 4, whether the burden of covering the costs is done in such a way to minimize undue reductions in energy use. Policymakers also should keep in mind the possibility that changes in the current regulatory system to generate extra revenues for stranded-cost recovery might well violate implicit contracts between the government and electricity customers.

As a practical matter, the amount of stranded-cost recovery, as well as the method for recovering it, is likely to be determined through a political bargain among the utilities, major customers, new entrants, and the public as represented by the regulators. The most crucial task facing regulators, then, is finding a way of dealing with stranded costs that does not sacrifice many of the benefits of competition. Economic principles, applied to regulation and contracting, should play a role in developing and evaluating solutions and helping the process to benefit both the electricity industry and the economy as a whole.

Chapter 7

Implications of Restructuring for Environmental Protection

Policymakers face many choices about whether and how to implement competition, regulate transmission grids, compensate for stranded costs, and restructure vertical relationships in the electricity industry. To this point, our focus in analyzing these choices has been the likely economic performance of electricity markets. In other words, we have viewed markets and competition as ways to deliver electricity to homes, businesses, and factories at the lowest possible cost. Research and experience suggest that competition generally best achieves economic efficiency, by making electricity available to anyone willing to cover the costs of producing it.

But considerations other than competition should and already do play a role in forming policies toward the electricity industry. For instance, leaving electricity provision entirely to the market may lead to a situation where the poor or less-advantaged members of society are relatively worse off. Policies to provide low-priced services to low-income customers, subsidized federal power systems, rural electrification, and other programs reflect such social concerns.

In this chapter, we devote our attention to another potential social concern—the need for environmental protection. Competition may not be efficient from the standpoint of society as a whole if prices do not reflect the environmental costs of electricity generation, transmission, and use as well as the more obvious costs associated with labor, equipment, and raw materials. When the costs of environmental damage associated with generation, transmission, or distribution are not borne by the company producing or delivering the power, we have what economists refer to as an *externality*. (For more about externalities, see box on page 111.) Externalities are said to be *positive* when the side effects are beneficial, and *negative* when the side effects cause harm. Environmental externalities associated with electricity are usually negative, although in many cases electricity may have side effects that are less harmful than those associated with alternative energy uses, such as wood stoves or industrial boilers.

Over the years, electricity use has been associated with a variety of negative externalities, including the following ones.

- During the 1970s, when a sizable fraction of peak power was produced from oil, there was much concern about growing U.S. dependence on imported oil and the effects of this dependence on national security.
- Fears associated with radiation from nuclear power facilities received considerable policy attention.
- Some scientists believe that harmful side effects may be caused by the electromagnetic fields that surround high-power transmission lines; others doubt the evidence.
- Power plants may affect lakes and rivers by emitting heavy metal elements and discharging hot water.

In addition, many people believe that increasing use of energy would deprive future generations of natural resources, as well as encourage drilling, mining, and other activities that degrade the environment.

The negative externality that receives the most attention from policymakers, however, and the one that is most likely to be affected by electricity deregulation and restructuring comes from the air pollution emitted by power plants as they burn fossil fuels to generate electricity. Because of the severe consequences of power plant emissions, the electricity industry has been subjected to many policy initiatives to reduce air pollution and its consequences. Whether this is relevant to the debate about restructuring the electricity industry depends on the answers to two questions. First, is the need for these environmental regulations likely to change? And, second, is the current spectrum of regulations likely to be more or less effective as a result of increased competition in the industry?

What is an externality?

An *externality* occurs when a cost or benefit can be identified and yet a price cannot be determined because a market does not (or cannot) exist. Because externalities have no price tags, they are not recorded as part of a company's private cost of producing an item, and so they are not reflected in the item's market price.

An externality is *positive* when the unmarketed effect is a benefit. Inoculation against a contagious disease is an example of a product or service generating a positive externality. Inoculation benefits the person inoculated (who does not get sick), but it benefits other people too (who cannot be infected). Though the benefit to others is evident, it is difficult to put a price on it.

An externality is *negative* when the unmarketed effect is a harm. Air pollution is the textbook example of a negative externality. There is typically no way to set up a market either to charge people who want to dirty the air or for people to be paid for not dirtying it. Emissions taxes and permit programs that require firms to purchase licenses to pollute are both policies that can simulate what would take place if we had conventional markets in rights to "use" clean air.

To explore these questions, we begin with a description of air pollution effects associated with electricity generation and the federal and state policies enacted in response. In the subsequent sections of this chapter, we identify many different ways in which changes in electricity markets resulting from industry restructuring can have an effect—either negative or positive—on the environment. At present, the high degree of uncertainty surrounding the likely effects of industry restructuring on the behavior of electricity producers, customers, and policymakers makes such prediction futile. Yet it is possible to offer some insights into the factors likely to have the greatest effect and to suggest the most productive avenues for future research.

Dealing with Air Pollution

Emissions from Electricity Generation

The generation of electricity produces at least four troublesome forms of air pollution, with the severity of effects on health and the environment depending on the concentration of the pollutant in the air, the length of time and intensity of exposure (for instance, for days rather than minutes, from inhalation while jogging rather than sleeping), and the sensitivity (of the lung or of a crop, for example) to the dose.

Particulate matter (PM)—soot, dust, dirt, aerosols—has readily apparent effects on visibility and exposed surfaces. In addition, PM can create or intensify breathing and heart problems and lead to cancer and premature death. Because the smaller particles are believed to cause the most damage (although the most potent size and composition are still uncertain), the U.S. Environmental Protection Agency (EPA) focuses on what it calls PM–10, which are those particles in the air smaller than ten microns in diameter (about 1/2,500 of an inch). In the future, it is quite possible this standard will be tightened to control even finer particles, down to 2.5 microns in diameter. While the health and visibility effects associated with particulates may be severe, controlling direct PM emissions from generators has proven to be inexpensive and effective.

Sulfur dioxide (SO_2) is a gas that may affect the heart and lungs in ways similar to particulates. In the air, some SO_2 (and nitrogen dioxide, described below) can convert into very fine particulates, which adds to ambient PM–10 concentrations. In addition, SO_2 may damage trees and lead to acid rain, which can harm lakes and streams and also corrode exposed materials, such as the outsides of buildings. Title IV of the 1990 Amendments to the Clean Air Act placed a nationwide cap on annual emissions of SO_2 and permitted electric utilities (and other SO_2 emitters) to buy and sell the rights to emit SO_2 within the limits of the cap.

Nitrogen dioxide (NO_2) is a brownish gas with the potential to cause adverse effects similar to those associated with SO_2 and may cause increases in PM–10 concentrations, particularly in the western United States. In the presence of sunlight and volatile organic compounds, NO_2 can contribute to the formation of ground-level ozone (or smog), which causes respiratory problems and crop losses. Ozone standards are violated in more areas than other air quality standards are.

Nitrogen dioxide is produced when nitric oxide (NO) emitted from power plants combines with oxygen already in the atmosphere. Many discussions of nitrogen-based air pollution refer to nitric oxide and nitrogen dioxide together as nitrogen oxides, or NO_x. Of the three electricity-related pollutants most responsible for current environmental damages—particulates, SO_2 and NO_x—nitrogen oxide emissions currently merit the most immediate policy attention. NO_x emissions are neither capped by environmental regulation, as are SO_2 emissions, nor controlled as effectively as particulates with current technology.

Greenhouse gases consist primarily of carbon dioxide (CO_2), which is a by-product of the burning of fossil fuels. Greenhouse gases are thought to contribute to *global warming,* a general increase in the temperature at the earth's surface. The extent of both the global warming effect itself and the harm it may cause continues to be controversial. If the damage from global climate change associ-

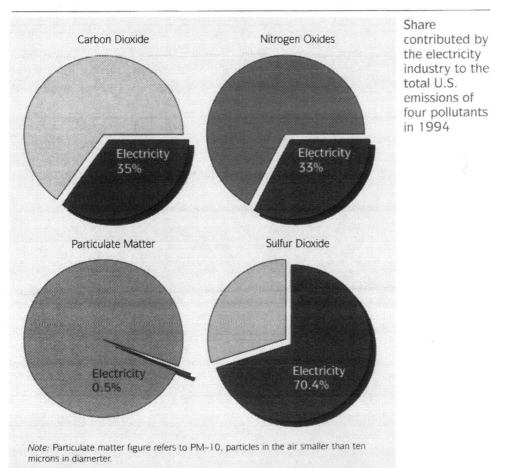

Carbon Dioxide

Nitrogen Oxides

Electricity
35%

Electricity
33%

Particulate Matter

Sulfur Dioxide

Electricity
0.5%

Electricity
70.4%

Share contributed by the electricity industry to the total U.S. emissions of four pollutants in 1994

Note: Particulate matter figure refers to PM–10, particles in the air smaller than ten microns in diamerter.

Source: U.S. Environmental Protection Agency, *EPA Air Quality Trends: Data Appendix,* 1995.

ated with greenhouse gas emissions ultimately proves to be large, any analysis that does not include them will be highly misleading.

In addition to these major pollutants, electricity generation also can lead to the emission into the air of toxic heavy-metal elements such as mercury, lead, or cadmium.

Power plant emissions and associated damages can vary greatly based on the amount of electricity produced, the time when it is produced, the fuel used, engineering efficiency and controls, and the plant's age and condition. For example, gas-fired units generally produce lower emissions of all air pollutants per kilowatt-hour than do coal-fired units. Under certain conditions, NO_x emitted from power plants actually can reduce ambient ozone concentrations, at least in areas near the smokestack.

> *Considerations other than competition should and already do play a role in electricity policymaking. One important concern is environmental protection, since the consequences of power plant emissions can be severe.*

In spite of far-reaching regulatory activity and major investments in energy efficiency and pollution abatement during the last two decades, emissions from power plants contribute substantially to the nation's NO_2 concentrations, greenhouse gas emissions, and SO_2 emissions. (The electricity industry's contribution to the total amount of these pollutants in the United States is shown in the figures on page 113.) The industry's SO_2 emissions, however, have already begun to fall dramatically as utilities are coming into compliance with the emissions reductions required by the 1990 Amendments to the Clean Air Act.

Damages from Air Pollution and Policy Options

Identifying harmful emissions is only the first step in assessing the environmental harm caused by electricity generation. Next, the emissions need to be measured and described as amounts of pollutants in the environment, called *ambient concentrations*. Ambient concentrations are meaningful only to the extent that people, ecosystems, and vulnerable objects are exposed to them. Therefore, the health, ecological, and other effects of *exposure* to these pollutants must be ascertained next. The last step is to calculate *damages*, in order to determine how much society might gain by reducing or eliminating those harmful effects.

Ideally, the cost of the associated environmental damage would be included in the cost of producing electricity. Adding these damages to the fuel, capital, and other costs of electricity yields what is known as the *social cost* of electricity. When the price of electricity is below its social cost, users will consume too much of it. Beyond some point, the value users place on the electricity they are purchasing is less than the social cost of producing that electricity, including externalities. In that sense, that electricity is not worth producing, despite the willingness of users to cover the direct costs of generating and distributing it.

From an economic perspective, the objective of environmental policy regarding the electricity industry is to figure out how to get consumers and producers to take these extra environmental costs into account. Forcing both consumers and producers to recognize these costs will encourage efforts either to reduce electricity use or to invest in cleaner generating technologies or pollution abatement equipment. To achieve these goals, policymakers make use of environmental policies that fall into two broad categories. *Command-and-control regulations* require that polluters meet specific emission-reduction targets and often require the installation and use of specific types of equipment to reduce emissions. Examples of this type of regulation include new source performance standards implemented after the 1977 Amendments to the Clean Air Act; these standards mandated smokestack scrubbers on new power plants that would remove harmful gases. *Incentive-based regulations* attempt to give polluters an economic reason to account for the damage that their pollution causes. This accounting may involve paying taxes for pollution or allowing firms to trade permits or licenses to emit pollutants. Other incentive-based (but generally less preferred) policies include the government's setting up subsidy programs that induce firms to reduce emissions or that encourage electricity users to install more fuel-efficient appliances.

Command-and-control mandates are most likely to be desirable in settings where there is clearly a single, most-economical way to meet environmental policy goals. However, most economists and many environmental policy advocates have come to prefer incentive-based approaches. The flexibility of incentive-based approaches allows polluters to find the least-expensive ways to reduce environmental damage. Such flexibility increases the amount of environmental protection that can be achieved for a given sum spent on abatement equipment or reduced electricity use.

How Restructuring Could Affect the Environment

From an economist's standpoint, competition in electricity markets is attractive because it should lead to lower electricity prices by encouraging the use of more-efficient, low-cost power plants.

Environmental effects of increased coal-fired generation after restructuring

Once greater access to the transmission system is allowed, the use of existing, coal-fired generators (especially in the Midwest) is expected to increase. The environmental effects of this increased use have been the subject of several studies, including an analysis by the Federal Energy Regulatory Commission (FERC) of the environmental impacts of its 1996 Order 888 on transmission access, a study by Henry Lee and Negeen Darani of Harvard University's Kennedy School of Government, and continuing work at Resources for the Future (RFF). While the methodologies and underlying assumptions vary across these different studies, each addresses the likely effects of greater transmission access on emissions of nitrogen oxides (NO_x) and the subsequent effects on air quality.

The FERC analysis compares the results of implementing the generic transmission access rule with the results of granting access to the transmission grid on a case-by-case basis and only when requested. This study finds that the rule will have small impacts on NO_x emissions from Midwestern generators and, therefore, on ozone concentrations in downwind areas. However, in its failure to consider the effects of restructuring on electricity demand and on the economics of nuclear generation, the FERC analysis may understate the environmental impacts of industry restructuring.

The study by Lee and Darani finds that restructuring could lead to substantial increases in NO_x emissions. They demonstrate that a 78 billion kilowatt-hour (kWh) increase in generation from existing coal facilities (two-thirds of which are assumed to be replacing generation from less-pol-

Lower prices, in turn, will lead to growth in the use of electricity. As the electricity industry changes, so will the technologies and fuels used to generate electricity and the size and location of plants. These changes, however, will have a direct bearing on the industry's airborne emissions and therefore on health and the environment.

Falling prices for electricity—and therefore increased production of it—need not necessarily spell increases in air pollution. For example, lower electricity prices could lead individual, commercial, and industrial users to substitute electricity for other energy sources that contribute more significantly to air pollution and environmental degradation. This could happen if, say, cheaper electricity led to increased use of electric-powered lawnmowers and automobiles in lieu of those equipped with gasoline-

luting facilities and one-third of which is assumed to be supplying new demand) could lead to a 500,000 ton increase in NO_x emissions. While this effect may be overstated because the authors assume that all 78 billion kWh comes from the most-polluting coal-fired generators, it also could be understated because the authors are conservative in their estimate of how much additional electricity existing coal-fired plants would generate to sell into competitive markets.

Ongoing research at RFF suggests that the amount of additional generation from existing coal-fired facilities that might be stimulated by restructuring could be substantially larger than Lee and Darani's estimate, depending in part on the rate of growth of transmission capacity. Regional differences in generation costs and electricity prices will be key determinants of which regions of the country will be power exporters and which regions will be power importers. The emissions and air quality effects of this power trading will vary by region and will depend on the relative locations of power exporters and importers and how the imported electricity is being used. In general, air quality effects will be less severe if the imported power is largely displacing generation by coal-fired generators in importing regions than if it is largely displacing nuclear generation.

Sources: Henry Lee and Negeen Darani, "Restructuring and the Environment," CSIA Discussion Paper 95-13, Kennedy School of Government, Harvard University, December 1995.

Federal Energy Regulatory Commission, *Promoting Wholesale Competition Through Open Access Non-Discriminatory Transmission Service by Public Utilities* (RM95-8-000) and *Recovery of Stranded Costs by Public Utilities and Transmitting Utilities* (RM94-7-001), Final Environmental Impact Statement, April 1996.

Dallas Burtraw and Karen Palmer, "Electricity Restructuring and Regional Air Pollution," Draft Working Paper, Resources for the Future, May 1996.

powered, internal combustion engines. Electric motors also could substitute for coal- or oil-fired generation in industrial processes.

Restructuring is likely to affect the mix of old and new plants generating electricity. In the medium term, industry analysts expect to see increased use of old plants. Currently, quite a few older, coal-fired generators in the Midwest are generating well below their capacity limits; these generators generally have operating costs well below current electricity prices in many parts of the country. The opening up of the transmission grid described in Chapters 3 and 4 will provide these Midwestern generators with increased access to distant markets, leading them to run more intensively. (Access to distant markets, however, still will be limited by the capabilities of the national transmission grid.) Recent and ongoing research suggests that increased generation from these Midwestern plants is likely to increase NO_x emissions. These increased NO_x emissions could contribute to air pollution problems in the form of higher concentrations of PM–10 and, possibly, higher concentrations of ground-level ozone in the Northeastern states and other regions downwind from these power plants. (For information about recent studies that analyze this issue, see the box on pages 116–117.) On the other hand, higher NO_x emissions in the Midwest may lead to local *decreases* in ozone concentrations, due to the complex chemistry of ozone formation.

In the long run, restructuring may lead to greater reliance on new plants with more benign environmental consequences. To the extent that restructuring leads to the use of new, cleaner plants (that, for instance, use combined-cycle gas turbines) instead of older ones (that, say, use coal), then reductions in damage from air pollution could be expected. Recently completed analyses of fuel-cycle costs estimate that the dollar value of the health and environmental damage associated with new coal, gas, and oil plants is only about 3–8 percent of the price of electricity—excluding the highly uncertain yet potentially large effect of greenhouse gases. In contrast, a typical older coal-fired plant emits from two to five times the nitrogen oxides and sulfur dioxide per kilowatt-hour as a new plant, though the latter is governed by a nationwide emission cap.

Beyond the relative age and fuel source of power plants, their physical characteristics, especially location, can contribute greatly to their effects on the environment. When access to transmission grids is improved, the share of electricity supplied by generators far from population centers may gradually increase, potentially reducing human exposure to pollutants and the associated harm to health and visibility—if, for instance, more electricity is generated downwind from population centers or near the ocean. When entry into the generation market is easier, however, new plants could be built closer to customers, which could increase the adverse health effects if the plants are upwind. These new plants are likely to be small, however, and would use less-polluting fuels (such as natural gas) and renewable fuels, rather than coal or oil. If so, this, too, could point in the direction of reduced damage from pollution.

Another physical characteristic of power plants—smokestack height—has an environmental effect. The higher smokestacks found in large, coal-fired power plants distribute emissions farther downwind and disperse them more widely than do the shorter stacks found on smaller units and on industrial cogeneration plants. Whether this effect helps or harms the environment depends on whether a small change in exposures experienced by many people provokes the same aggregate response as a large change experienced by proportionally fewer people. If the aggregate response to the small change is less than to the large change, then tall stacks could be environmentally beneficial. Said another way, if restructuring leads to more small units and more cogeneration, especially near populated areas, the damage from ambient concentrations of pollutants might increase. However, lower prices could lead industry to buy more power from generation companies and cogenerate less, reducing concentrations and exposures.

Restructuring and Environmental Policies

Whether or not competition in electricity generation markets affects air pollution, it certainly will affect the many policies regulators have devised to deal with power plant emissions. The U.S. electricity industry has been the object of many different types of federal environmental regulation since the passage of the Clean Air Act in 1970, and the federal regulatory reliance on incentive-based methods, such as tradable sulfur-dioxide emission allowances, has been growing. At the state level, environmental concerns have led some public utility commissions (PUCs) to require their utilities to take the social costs of pollution into account in planning the construction and operation of power plants. In addition, some state governments have required utilities to initiate demand-side management (DSM) programs aimed at conserving energy; some state governments also have adopted policies requiring or favoring use of renewable or inexhaustible energy resources, such as wind and solar energy. In the last half of this chapter, we examine more closely each of these regulatory options.

Marketable Emission Allowances versus Command-and-Control Regulation

One of the leading policy innovations in recent years has been the 1990 legislation and subsequent regulations creating EPA's Acid Rain Program—a system of marketable emission allowances instituted to control sulfur dioxide. Under this program, Congress set a national cap on the total amount of SO_2 that polluters can release into the atmosphere; ultimately, the cap will reduce total annual emissions by nearly 50 percent from 1980 levels. Congress also directed EPA to allocate SO_2 emission allowances to coal-burning power plants based on proportionate reductions from past rates of emissions. A generator that emits less SO_2 than it has allowances for can sell the excess allowances to other utilities that find it less costly to buy emission allowances than to install additional abatement equipment, to switch to less-polluting fuels or technologies, or to forgo additional electricity sales.

The resulting market for allowances creates a price that anyone needing to discharge SO_2 can pay. A utility that wants to emit more SO_2 than it holds allowances for must buy additional allowances from another utility that reduces its emissions. Equally important, a utility that reduces its emissions is rewarded by virtue of having an extra allowance to sell. In both cases, the program forces utilities to factor the cost of additional pollution into their production and technology decisions, in that they have either to buy more allowances or use some of the allowances they already hold and could sell. This policy directly addresses the chief cause of the pollution problem—that in the past polluters and their customers have not had to take environmental damage into account—and it does so in a way that ensures that the total cost of controlling SO_2 is minimized.

For marketable allowance programs to accomplish their purpose, however, generation companies holding emissions allowances must be interested in using them in the most profitable way—that is, generators must have a real incentive to lower their costs. Under traditional rate-of-return regulation, where prices reflect average costs, a utility may lack this incentive. For example, suppose state PUCs force utilities to pass along to their customers, in the form of lower prices, any profits from installing SO_2 abatement equipment and then selling excess allowances. If so, the utilities will not have much interest in selling allowances or reducing pollution. (This description characterizes the approach used to date in every state except Connecticut.) However, in a deregulated market, where generators are free to earn as much as they can, the incentive to make these trades and to install low-cost abatement technology or switch to cleaner fuels will become stronger. Therefore, we expect that in a competitive electricity market the advantages of emission allowance trading programs relative to command-and-control methods of regulation would be greater than in a regulated market.

Social Costing versus Taxation

Social costing refers to several different methods by which state regulators induce utilities to factor air pollution damage into their business decisions. One of those methods, called *social-cost planning* has been undertaken in seven states. Social-cost planning involves estimating *adders,* which represent the environmental damage from particular generation technologies, and adding them to the private cost of producing electricity with these technologies. Regulators then use this sum as a proxy for the social cost of different modes of electricity generation and require utilities to base their investment (or contracting decisions, in the case of power purchases from nonutility generators) on those social costs.

Although social-cost planning has been highly contentious, the relatively modest nature of the adders relative to the cost of generation has not led to a dramatic reordering of investment plans. From a practical standpoint, however, retail competition in the electricity industry would make social-cost planning virtually unworkable. New electricity suppliers, while subject to environmental standards imposed by EPA, would be able to avoid social-costing regulation in the planning process by operating as independent wholesale generators and marketing power directly to large customers or to power aggregators who then contract directly with electricity consumers. If incumbent utilities are no longer the primary suppliers or wholesale purchasers of generated electricity from new generators, then regulation of either their investment in generation or their contracting practices with new generators will become irrelevant.

Social costing induces utilities to factor air pollution damage into business decisions. But if social costing is applied only to the electricity industry, consumers may shift their purchases to more-polluting fuels and technologies.

A different version of social costing entails a *social-cost dispatch system*, in which electricity would be dispatched according to its social, rather than its private, cost. Under a social-cost dispatch system, generators would provide the system dispatcher with both a price bid and information (such as emission rates, stack heights, and location coordinates) necessary to calculate the external costs of operating the plant. A social-cost dispatch system has an advantage over social-cost planning in that it covers both old and new generators. By imposing adders only on new plants, social-cost planning would encourage utilities to continue using older, dirtier units. In contrast, social-cost dispatch contains no such bias. Social-cost dispatch also would permit regulators to change the use of all generating technologies in response to atmospheric conditions. For instance, on windy days, it might make sense to allow expanded operation of cheaper but dirtier plants; during heat inversions, when the air is stagnant, more-expensive, cleaner plants could get heavier use.

Under competition implemented via bilateral contracting (described in Chapter 3), social-cost dispatch may be untenable, however, because generators could cut independent deals with distributors or customers beyond the reach of regulators. Without a central dispatcher, there is no one to calculate damages and schedule generating units to minimize social costs. If competition were implemented via the Poolco model (also described in Chapter 3), then all power could be dispatched on the basis of social costs through a centralized Poolco. Implementation of this method requires that the pool be mandatory since, under a voluntary pool, any unit wishing to avoid having its external costs considered in the dispatch could merely opt out of the pool.

Since the Poolco approach would be necessary to make social-cost dispatch effective, the benefits of social costing should be weighed against the disadvantages of the Poolco system and the benefits of generators' having direct access via transmission services to their customers. In addition, if social-cost dispatch leads to higher electricity rates because higher-cost but less-polluting generators are favored, consumers may shift their consumption away from electricity and toward other fuels with prices that do not reflect *their* pollution costs.

If such a shift becomes a problem, then the obvious solution is to abandon social costing for electric utilities alone and impose appropriate emission taxes throughout the economy to force everyone to incorporate the cost of environmental damage. Why concentrate on electric utilities alone, especially when doing so might encourage a switch to more-polluting fuels and technologies? If all pollution were taxed, the choice to adopt bilateral contracting or Poolcos could be decided on their merits, without having to factor in any need for environmental adjustments through Poolco dispatch decisions. Indeed, restructuring could make it more likely that such taxes would lead to efficient emission reductions, because generators and distributors would have stronger incentives to control costs. Under conventional rate-of-return regulation, such taxes simply might be passed along to ratepayers without influencing investment or dispatch decisions. If political considerations dictate that such taxes are unlikely to be adopted, however, then social-cost dispatch may be desirable as long as the reduction in social costs associated with electricity supply is not swamped by other harmful effects.

Demand-Side Management

Demand-side management (DSM) refers to a variety of energy conservation programs adopted by utilities and intended to address the problems (including environmental damage) associated with increased electricity use by encouraging customers to reduce their electricity consumption. DSM programs include free energy audits, subsidies for the purchase of energy-efficient heating and air conditioning systems, and installation of devices to limit or interrupt loads. The vast majority of electricity customers are served by utilities that offer DSM programs. In the United States in 1992, utility expenditures on DSM programs were $2.4 billion, up from $0.9 billion in 1989.

It remains controversial whether, as a method for reducing pollution, DSM is more cost-effective than emission allowance trading programs, social-cost dispatch, or pollution taxes. Proponents of DSM argue that consumers are unaware just how much they could save through energy conservation and hence benefit from a utility's adoption of such programs. Skeptics challenge such evidence on the grounds that it ignores differences in performance among types of lighting (for instance, fluorescent lights may be more energy-efficient, but they may be less pleasing or harder on the eyes than incandescent lights), variability of fuel prices, and other factors important to consumers.

A sounder claim, perhaps, in favor of DSM is that it encourages conservation when electricity prices should rise but do not—for instance, at peak demand periods, such as hot summer afternoons. This claim for DSM programs loses force, however, if deregulating electricity enhances the ability of electricity suppliers to base prices on time and load conditions.

Despite the controversial nature of DSM programs, many state PUC commissioners and other government officials insist that DSM efforts be maintained if restructuring is to move forward. These officials believe that these programs facilitate cost-effective energy conservation, of which customers often fail to take advantage on their own. However, in some states, demand-side management programs are being cut back. For example, the Michigan legislature recently eliminated DSM requirements on the grounds that they discriminate against ratepayers unwilling or ineligible to participate in DSM programs—such as those not in a financial position to install a new high-efficiency air conditioner, even taking a DSM subsidy into account.

If DSM subsidy programs do indeed continue, then the problem becomes finding a way to fund them in a restructured industry where competition may prevent regulators from requiring that incumbent utilities cover DSM costs. With technology making profitable self-generation available to smaller and smaller firms, imposing DSM costs on utilities may encourage self-generators to bypass the utilities' transmission or distribution systems in significant numbers. The National Resources Defense Council has offered a solution in the form of a "system benefits charge," wherein state PUCs levy a nonbypassable charge to prevent customers from leaving incumbent utilities solely to avoid paying the costs of DSM programs.

Renewable Energy Sources

Renewable resources are energy sources that do not use exhaustible resources as fuels. Sources of renewable energy include water, wind, solar energy, and geothermal energy, as well as some combustible materials, such as landfill gas, biomass, and municipal solid waste. Water is the leading renewable resource when it comes to electricity production, with hydroelectric plants generating about 12 percent of all U.S. electricity in 1994; the other renewables made up an additional 2 percent.

Although renewable energy generally is considered to be environmentally friendly, it brings its own environmental problems. In the case of hydroelectric facilities, for example, the damming of watercourses can destroy or damage natural habitats, as well as adversely affect recreational opportunities. During the next fifteen years, a quarter of all hydroelectric capacity in the United States will come up for relicensing by the federal government. The relicensing process, in turn, will require a reexamination of the use of water resources in energy production, including mitigation of environmental effects. Some relicensing decisions may hinge on utilities' making expensive commitments to improve facilities and to accommodate recreation and seasonal fish migrations. The cost of these commitments may drive some current providers of hydroelectric power away from the business because their projects are no longer profitable; this possibility is exacerbated if generation markets are opened to competition.

Federal and state policies have boosted the use of renewable resources in electricity generation. Restructuring may call for new approaches to energy conservation and environmental protection.

The use of the other renewable resources has been boosted by federal law and state policies, as we described in Chapter 2. In 1978, the Public Utilities Regulatory Policies Act (PURPA) mandated that electric utilities purchase power from qualifying facilities at the utilities' own avoided cost of production. California and some other states then set avoided costs at high rates. Together with rising prices for natural gas and oil in the late 1970s and early 1980s, this practice encouraged a surge in supply of generators using renewable technologies. However, in response to declining natural gas prices in the early 1990s, state regulators lowered the avoided-cost payments. As a result, the growth in generation capacity relying on renewable resources slowed from an annual rate of 9.7 percent in 1989 to 1.2 percent in 1993. In 1992, the Energy Policy Act (EPAct) gave some renewables a boost by making permanent a 10 percent energy tax credit for solar and geothermal projects. EPAct also established a Renewable Electricity Production Credit, which grants a 1.5 cent/kilowatt-hour, ten-year tax credit for electricity generated using new wind facilities and some types of biomass technologies.

Restructuring places at risk renewable energy sources for electricity generation. In an unregulated environment, strategic planning by private firms will tend to replace public utility planning subject to PUC oversight. Furthermore, as the cost of electricity generation falls, so too will the estimates of avoided costs that are used to compute the prices paid to generators using renewable resources, lowering their earnings. In some cases, the prices that renewable generators can charge will no longer cover their costs, largely vitiating the PURPA qualifying-facility program. However, some analysts have suggested that, with proper pricing of transmission, the advantages of localized generation will be recognized; changing the mix of generation facilities from large generators serving wide areas to smaller, more localized plants will lead to a greater use of some forms of renewable technologies. In addition, through bilateral contracting, industrial and commercial users could take advantage of wind and solar generation during periods of peak power needs.

Whether environmental costs from fossil fuels should lead us to grant special treatment for renewable resources (which emit few, if any, greenhouse gases) is likely to depend on global warming and the extent to which renewable technologies would replace older, dirty coal-fired plants. Moreover, existing government policies may be working against renewables. Companies using renewable resources may be paying twice as much in taxes as those using fossil-fuel technologies, because capital and construction are taxed heavily relative to operating and maintenance expenses. Expanding and increasing EPAct's investment tax credits for renewable electricity generation, while eliminating all other preferences for renewables, may be an appropriate policy response in a restructured electricity industry.

New Policies To Address the Environmental Consequences of Restructuring

Not only will restructuring have an effect on existing environmental policies, but it also may require new policies to address its environmental consequences. As we already noted, the long-run environmental consequences of restructuring are difficult to predict today. However, industry observers generally agree that allowing greater access to the transmission grid is likely to increase generation and, therefore, emissions of NO_x and CO_2 from existing Midwestern coal-fired generators, especially in the medium term. Concern about these increased emissions has been the major focus of the debate regarding the environmental effects of the Federal Energy Regulatory Commission's (FERC) Order 888 (issued April 1996) on transmission access and the need for new environmental policies to mitigate these effects.

The appropriate policy response to the increase in NOx emissions from Midwestern generators could be to build on new environmental policies currently being developed by EPA and state governments. To reduce the contribution of NO_x emissions within the Northeast to ground-level ozone problems in that region, state governments of all of the states from Maryland and the District of Columbia north and as far west as Pennsylvania, in cooperation with the EPA, are developing a NO_x emission allowance trading program. Like the sulfur dioxide trading programs discussed earlier in this chapter, this program would involve trading of NO_x emission allowances among utilities and large industrial sources of NO_x emissions throughout the multistate region. This program should enable these Northeastern states to reach the NO_x emission cap for the entire region at minimum cost, given the boundaries of the trading areas. If NO_x emissions from Midwestern states are contributing to ozone and other air pollution problems in the Northeast, then the region covered by this trading program should be expanded to include utility and other sources in the Midwestern states as well.

Establishment of this proposed NO_x emission allowance trading program is likely to take time, and it may not be in place by the time the transmission grid is opened up and access to competitive electricity markets is expanded. Therefore, interim measures may be necessary to soften the environmental effects of open transmission access while more comprehensive policy evolves at EPA. One possible approach might include a requirement that generators increasing production for transmission to distant markets mitigate the associated increase in emissions either by adding pollution controls at their generators or by paying nearby facilities to reduce their NO_x emissions—so-called *emissions offsets*. However, any industry-specific approach would be and should be eclipsed if a more comprehensive program can be implemented by EPA that would permit cost savings through interstate and interindustry emission allowance trading.

Conclusion

In the absence of environmental regulations, introducing competition into electric power generation could either exacerbate or reduce environmental damage. On the one hand, reduced prices would increase use and emissions, particularly in the short to medium term when much of the increased generation is expected to come from existing coal-fired generating plants. However, in the long run, competition is likely to invite market entry by new generation companies making use of newer, cleaner technologies and, perhaps, displacing power supplies from older, less-efficient, and dirtier plants.

The continuing presence of a federal and state structure for environmental regulation suggests that restructuring may have limited implications for the environment. Cities will still have to meet air quality standards, power plants will still be subject to an SO_2 emissions cap, and policy initiatives for preserving air quality in national parks and rural areas will still continue. Indeed, public policies designed to take advantage of the market forces unleashed by restructuring may further improve the environment, promote conservation, and result in a more-efficient level and mix of power supplies. While state policies directed at the environment, including DSM programs and social-costing activities, are likely to diminish or even end, current PUC efforts to tilt utility investment decisions to be favorable toward the environment have not been widespread or particularly forceful, so little will be lost.

Nevertheless, before anyone can be optimistic about the effects of restructuring on the environment, gaps in current federal and regional air pollution control policies will need to be filled. For example, if the efforts in the Northeastern United States to cap nitrogen-oxide emissions within a tradable allowance system are successful, the need for concern about this pollutant would be diminished; this is particularly true if the trading area expands to include Midwestern states, which promise to see an increase in generation activity and associated nitrogen-oxide emissions in a more competitive environment. A competitive industry likely would be more responsive to emission allowance trading programs and other incentive policies than are utilities that have been insulated from profit and loss under traditional forms of regulation.

Where change is more certain is in the pace and character of conservation efforts and renewable energy use. PUC directives were much more widespread and effective in this area than in the environmental area. Unless the federal or state governments pick up the slack, certain subsidized forms of conservation and renewables may become less effective when the electricity industry is restructured. Whether society is the worse for these changes remains, like so many issues associated with restructuring, a subject for further study.

Epilogue

We hope that our review of the history and regulation of the electricity industry, the models of implementing competition, the regulation and technology of transmission networks, vertical integration, the stranded-cost controversy, and the range of potential environmental effects of restructuring will help bring some clarity and deliberation to a highly confusing and contentious debate. The issues that we address in this book cover the central issues raised by the movement to expand competition in the electricity industry. However, our portrayal cannot do full justice to the restructuring controversy without a brief mention of some other issues that also may come to play important roles as this drama develops in the years to come.

129

Merger Policy

As the debate over bringing competition to the industry intensifies, the number of utilities proposing mergers is growing. Announcements of merger plans often cite large cost savings that will be achieved as a result of these unifications. Skeptics suggest that other motives may be at work, such as attempts to monopolize regional electricity markets. In light of all of the uncertainty about future market conditions, Federal Energy Regulatory Commission (FERC), U.S. Department of Justice, and Federal Trade Commission officials responsible for evaluating these mergers have the difficult tasks of defining relevant markets and assessing future market concentration both with and without the proposed merger. The extent to which antitrust laws will affect how utilities compete—and coordinate—with each other will depend critically on how restructuring plays out in state and federal policy arenas. In addition, it is uncertain whether current laws and regulations allow authorities to review mergers properly before the policies that determine who competes with whom are worked out.

Entry into Telecommunications Markets

While a primary issue for electricity restructuring is how much the historical vertical integration of utilities should change, elsewhere are policy initiatives that could lead to expansion by electric utilities into other markets, most notably local telephone and cable service. The Telecommunications Act of 1996 amends the Public Utility Holding Company Act of 1935 to permit utilities to offer telecommunications services through subsidiaries, subject to detailed restrictions on financial arrangements and obligations to provide regulators with access to accounts and other operational information. Such entry may lead to greater competition in the local delivery of voice, data, and video, but also it also raises concerns regarding the discrimination and cross-subsidization that we discussed in Chapter 5. Whether the benefits of such entry by electric utilities exceed the expected risks of anticompetitive abuses and how the 1996 Telecommunications Act's rules compare with FERC's open-access regulations in protecting against similar abuses between transmission and generation could become important policy issues. Depending upon how enthusiastically utilities respond to the invitation to compete in telephone and cable service, policymakers dealing with the electricity industry may have to become familiar with an equally complex set of controversies that have dominated communications businesses for decades.

Electricity Research and Development

Much of the current research and development activity in the utility industry is conducted by the Electric Power Research Institute (EPRI), which is jointly funded by voluntary contributions from member utilities. Under the traditional approach to rate regulation, utilities have an incentive to participate in EPRI, since they are able to recover the costs of their contributions in the regulated utility rates paid by their customers. As competition puts pressure on utilities to reduce costs, many are expected to forgo participation in EPRI, choosing instead to selectively fund their own research projects as opportunities arise. However, research can be costly and, often, discoveries are easily appropriated by others, so that it is difficult to profit from successful research efforts. Many industry observers fear that, without an institution that performs research on behalf of many, the amount of research in the industry will diminish to the ultimate detriment of society. Other analysts argue that other institutions—such as the patent system and trade secrets designed to protect property rights to new discoveries—will provide sufficient incentive for R&D in the future.

Social Programs

As a result of its long history of economic regulation, the electricity industry has been forced to undertake many programs that it probably would not choose to undertake in a competitive world. Such programs include discounted rates for low-income consumers and economic development rates designed to help attract or keep businesses in economically disadvantaged areas of a state. As utilities face increased competition from nonutility generators or utilities from other states, the future of these programs becomes uncertain. Some observers have suggested that the costs of these programs can be recovered in regulated transmission and distribution rates; however, this approach may not always be feasible or desirable. If society wishes to continue these programs, alternative means may need to be found.

Utility Tax Policy

Federal, state, and local tax policies may have important implications for the functioning of competitive electricity markets. Across the country, many states, counties, and municipalities impose special taxes (including gross revenue taxes, franchise fees, and various tax surcharges) on utilities that are not imposed on other businesses. These taxes can be quite substantial, in some cases amounting to nearly 20 percent of utility revenues, so that utilities can make large contributions to public sector revenues. The prospect of bringing competition to the electricity industry could threaten those revenues as nonutility generators gain an increasing share of the electricity market.

At the federal level, utilities again may face an unequal tax burden, due in part to restrictions that limit their ability to rely on debt financing. These additional tax burdens may put utilities at a cost disadvantage relative to nonutility generators when they compete with each other in future electricity markets. In addition, investor-owned utilities also may be at a disadvantage relative to publicly owned utilities, which pay no income tax and can raise capital through tax-free bonds.

* * *

Those in both the private and public sectors with an interest in the evolution of the electricity industry—that is, just about all of us—have to face the prospect that restructuring problems are likely to become more numerous, chaotic, and complex before clear and simple solutions emerge. We hope that standing back to gain perspective on the major issues—as we have attempted to do with this brief look at some of the issues that may become prominent in the future—contributes to the development of policies that will promote the efficiency of the electricity industry and the well-being of the public.

Index

Printed in the United States
by Baker & Taylor Publisher Services